MATHEMATICAL
AMAZEMENTS AND SURPRISES

FASCINATING FIGURES
AND NOTEWORTHY NUMBERS

ALFRED S. POSAMENTIER AND INGMAR LEHMANN

AFTERWORD BY
HERBERT A. HAUPTMAN,
NOBEL LAUREATE

 Prometheus Books

59 John Glenn Drive
Amherst, New York 14228–2119

Published 2009 by Prometheus Books

Inquiries should be addressed to
Prometheus Books
59 John Glenn Drive
Amherst, New York 14228–2119
VOICE: 716–691–0133, ext. 210
FAX: 716–691–0137
WWW.PROMETHEUSBOOKS.COM

13 12 11 10 09   5 4 3 2 1

Library of Congress Cataloging-in-Publication Data

Posamentier, Alfred S.
    Mathematical amazements and surprises: fascinating figures and noteworthy numbers / by Alfred S. Posamentier and Ingmar Lehmann.
        p. cm.
    Includes bibliographical references and index.
    ISBN 978-1-59102-723-2 (pbk.:alk. paper)
    1. Mathematics—Miscellanea. 2. Number theory—Miscellanea. I. Lehmann, Ingmar. II. Title.
QA99.P665 2009
510—dc22   2009010091

Printed in the United States of America on acid-free paper

Contents

PREFACE

Too often mathematics is looked upon as a sequential arrangement of topics that seems to lead nowhere useful. Every once in a while, we are given some applications where the mathematics learned in school comes in handy, such as calculating the shortest path between two locations or determining the best price for various quantities of the same product. Of course, for those who work in a technical setting—science, engineering, economics, just to name a few—there is a constant need to use mathematics. Most people's introduction to mathematics has been bereft of exhibiting its beauty through the many unusual relationships within and between common topics. This book is designed to open up the world of mathematical wonders through largely numerical and geometrical phenomena. Naturally, there will be times when we will consider topics just a bit beyond simple arithmetic, such as some probability surprises, or use some algebra to justify some of these oddities. Yet throughout we will be cognizant of the general readership—keeping the language and the nomenclature simple so that it can be easily understood.

For example, we will consider one of the most counterintuitive illustrations in mathematics—in this case using probability. What do you think the likelihood is of finding two people in a room of thirty-five people who share the same birth date (not necessarily the same year)? Our intuition would dictate that it is quite unlikely, or the probability is rather low. Well, brace yourself; the actual probability is somewhat more than 80 percent! This is just one of the many surprises in store for you in this book.

There are many entertaining shortcuts in arithmetic or in recognizing number relationships at a glance, such as visually determining by what numbers a given number may be divided exactly. There are lots of unusual properties of specific numbers in our number system—many of which can be quite engaging, besides being useful. Exploring these will pique your curiosity and improve you insights into the nature of numbers. Here again, we will be gently building your appreciation for the beauty of mathematics.

When one speaks of beauty in mathematics, one might expect it to be visual. Of course, geometry lends itself quite nicely to that approach. There are many situations in which you begin with a rather general geometric shape and, by following simple consistent procedures, you end up—quite unexpectedly—with a beautiful geometric shape. For example, if you draw any (ugly) quadrilateral and join the midpoints of the sides, consecutively, with line segments, you will *always* end up with a parallelogram—sometimes a rhombus, other times, a square or a rectangle. The beauty that geometry provides for mathematics is boundless, so we present only some of the highlights for you to consider—mostly without proof that they are, in fact, true. Instead of just taking our word for their truth, we urge you to try to prove the relationships true. For this we provide you with an extensive bibliography. Beyond the beauty of the relationships lies the true beauty of the proofs that establish these amazing geometric marvels.

In short, this book will take you through a plethora of amazing and surprising examples of mathematical phenomena, each of which will contribute to your accepting our premise that there is real beauty—both visual and intellectual—in mathematics to be savored.

Chapter 1

AMAZING NUMBER PROPERTIES
AND RELATIONSHIPS

For the most part, numbers play a major role in our lives, allowing us to quantify and to create order. Yet numbers can be seen also for their own sake. They may harbor unusual properties, exhibit some hidden beauty, or just allow us to marvel at their inherent nature. In this chapter, we will investigate some of the attractive features that numbers can offer. We hope to amaze you with lots of unusual surprises.

There have been many ways to write numbers over the past millennia. The Egyptians, for example, used hieroglyphics to express numbers, and the Romans had a number system that is still used today, such as for chapter numbers or cornerstone dates. Yet these have proved to be far too cumbersome and inconvenient, and so when Leonardo of Pisa—better known today as Fibonacci (ca. 1175–1240)—introduced the "Indian numerals" 9, 8, 7, 6, 5, 4, 3, 2, 1, and 0 in the first words of his monumental book *Liber Abaci*, written in 1202, we had the use of these numerals for the first time in Western civilization. As we investigate the properties of numbers, we will use this standard base-10 notation.

In this chapter, we will introduce you to many beautiful number arrangements, number properties, number relationships, and interesting processes that we can use to analyze numbers. We will occasionally show a pattern and, in the absence of proof, ask you to accept the results, since we do not want to break the flow of the book with proofs. Yet, by the same token, we want you to be aware that not all "apparent patterns" of number relationships hold true for all cases. To give you just one illustration of a number pattern that appears to lead to a general result—but does not—we will consider the question of the French mathematician Alphonse de Polignac's (1817–1890) Conjecture:

9

**Every odd number greater than 1 can be expressed
as the sum of a power of 2 and a prime number.**

If we inspect the first few cases, we find that this appears to be a true statement. However, as you will see from the list in figure 1-1, it holds true for the odd numbers from 3 through 125 and then is not true for 127, after which it continues to hold true again for a while.

$$3 = 2^0 + 2$$
$$5 = 2^1 + 3$$
$$7 = 2^2 + 3$$
$$9 = 2^2 + 5$$
$$11 = 2^3 + 3$$
$$13 = 2^3 + 5$$
$$15 = 2^3 + 7$$
$$17 = 2^2 + 13$$
$$19 = 2^4 + 3$$
$$\vdots$$
$$51 = 2^5 + 19$$
$$\vdots$$
$$125 = 2^6 + 61$$
$$127 = ?$$
$$129 = 2^5 + 97$$
$$131 = 2^7 + 3$$

Figure 1-1

Perhaps you can find the next number that fails de Polignac's Conjecture. Remember, though, when we present you with a pattern, we will assure you that it will hold true for *all* cases.

In 1849, Alphonse de Polignac proposed another conjecture that has not been proved or disproved to date. It is as follows:

There are infinitely many cases of two consecutive prime numbers with a difference of some even number *n*.

For example, suppose we let $n = 2$. There are consecutive prime number pairs whose difference is 2, such as (3, 5), (11, 13), (17, 19), and so on. Note, we still have not established if this conjecture is true or false.

THE AMAZING ZERO SUMS

When you use a calculator to determine the following sum, you will find it to be zero.

$$123,789^2 + 561,945^2 + 642,864^2 - 242,868^2 - 761,943^2 - 323,787^2 = 0$$

This may be nice, since we have the squares of large numbers and they seem to show no particular pattern. Yet when we begin to manipulate these numbers in a very orderly fashion, the zero sum amazingly remains.

1. In the first case, we will delete the hundred-thousands place (the left-most digit) from each number:

$23,789^2 + 61,945^2 + 42,864^2 - 42,868^2 - 61,943^2 - 23,787^2$, and the sum remains 0.

We repeat this process by deleting the left-most digit of each number from each of the next few and look at the results:

$$3,789^2 + 1,945^2 + 2,864^2 - 2,868^2 - 1,943^2 - 3,787^2 = 0$$
$$789^2 + 945^2 + 864^2 - 868^2 - 943^2 - 787^2 = 0$$
$$89^2 + 45^2 + 64^2 - 68^2 - 43^2 - 87^2 = 0$$
$$9^2 + 5^2 + 4^2 - 8^2 - 3^2 - 7^2 = 0$$

2. We now will follow an analogous process, but this time we'll delete the units digit (the right-most digit) from each of the numbers, and again amazingly we see each time that the resulting sum is zero:

$$123,789^2 + 561,945^2 + 642,864^2 - 242,868^2 - 761,943^2 - 323,787^2 = 0$$
$$12,378^2 + 56,194^2 + 64,286^2 - 24,286^2 - 76,194^2 - 32,378^2 = 0$$

$$1{,}237^2 + 5{,}619^2 + 6{,}428^2 - 2{,}428^2 - 7{,}619^2 - 3{,}237^2 = 0$$
$$123^2 + 561^2 + 642^2 - 242^2 - 761^2 - 323^2 = 0$$
$$12^2 + 56^2 + 64^2 - 24^2 - 76^2 - 32^2 = 0$$
$$1^2 + 5^2 + 6^2 - 2^2 - 7^2 - 3^2 = 0$$

3. We will now combine the two types of deletions into one by removing the right and left digits from each number and, yes, again we retain zero sums!

$$123{,}789^2 + 561{,}945^2 + 642{,}864^2 - 242{,}868^2 - 761{,}943^2 - 323{,}787^2 = 0$$
$$2{,}378^2 + 6{,}194^2 + 4{,}286^2 - 4{,}286^2 - 6{,}194^2 - 2{,}378^2 = 0$$
$$37^2 + 19^2 + 28^2 - 28^2 - 19^2 - 37^2 = 0$$

This is not to be taken for granted; it is a truly amazing number relationship.

A MOST UNUSUAL NUMBER

In this book, we are not concerned about a number's mystical connections; our interest in numbers is purely mathematical. Yet in some societies for curious reasons some numbers symbolize good luck and others bad luck. For example, an inordinate number of children were born in China on August 8, 2008, and over seventeen thousand couples were married in Beijing on that date. It was also the date on which the twenty-ninth Summer Olympics opened in Beijing at 8:08:08 PM. Why that date? For the Chinese, 8 is a lucky number, and that date can be written as 08-08-08. What makes 8 so special is open for interpretation. Mathematically speaking, 8 is a perfect cube: $8 = 2^3$; and 8 is the only cube that is one less than a perfect square, 9. Also, 8, the sixth Fibonacci number, is the only Fibonacci number[1] (other than 1) that is a perfect cube.

On the other hand, the number 666, popularly known through its biblical association, is often referred to as the "number of the beast" and signifies bad luck omens. Yet again, we are only concerned about this number's mathematical properties, many of which are quite amazing. To begin with, the number 666 is obviously a palindrome[2]—that is, it is a number that reads the same in both directions. Yet if we were to write this number in Roman numerals—666 = DCLXVI[3]—we find that all the numerals less than 1,000 are used and in descending order!

1. The Fibonacci numbers are 1, 1, 2, 3, 5, 8, 13, 21, 34, 55, . . . , where each succeeding term is the sum of the previous two. See A. S. Posamentier and I. Lehmann, *The Fabulous Fibonacci Numbers* (Amherst, NY: Prometheus Books, 2007).
2. Palindromic numbers will be presented beginning on page 16.
3. D = 500, C = 100, L = 50, X = 10, V = 5, I = 1.

The number 666 just happens to be the sum of the first thirty-six numbers:

$$1 + 2 + 3 + 4 + 5 + 6 + 7 + 8 + 9 + 10 + 11 + 12 + 13 + 14 + 15 + \ldots + 30 + 31 +$$
$$32 + 33 + 34 + 35 + 36 = 666$$

Not only is the sum 666, but since the sum of initial consecutive natural numbers[4] always yields a triangular number,[5] 666 is a triangular number. Surely, we know that 36 is a square number (that is, 6^2). Therefore, for 666, we can say that a square number of initial natural numbers has given us a triangular number.

To further demonstrate the uniqueness of the number 666, consider the first seven prime numbers:[6] 2, 3, 5, 7, 11, 13, 17. If we take the square of each of them and then find their sum, yes, we arrive at 666.

$$2^2 + 3^2 + 5^2 + 7^2 + 11^2 + 13^2 + 17^2 = 4 + 9 + 25 + 49 + 121 + 169 + 289 = 666$$

Another amazing fact about this strange number, 666, is that the sum of its digits $(6 + 6 + 6)$ is equal to the sum of the digits of its prime factors. That is, since we have in prime factored form $666 = 2 \cdot 3 \cdot 3 \cdot 37$, the sum of the digits, $2 + 3 + 3 + 7$, is equal to $6 + 6 + 6$.

It is also curious that the sum of two consecutive palindromic prime numbers, $313 + 353$, is equal to 666.

The number 666 is equal to the sum of the digits of its 47th power, and is also equal to the sum of the digits of its 51st power. That is,

$666^{47} =$ 50499696844207967531731487984055647729415162952654081881176 3266 8936540446616033068653028889892718859670297563286219594665 90473 3945856

$666^{51} =$ 99354075759138594033426351134129598072385863746943100899712 0691 31346071328296758253023455821491848096074897283890063763421 5694 097683599029436416

You can check this by finding that the sum of the digits of each of the above large numbers is equal to 666.

---

4. Natural numbers are simply our counting numbers: 1, 2, 3, 4, 5, ... Sometimes 0 is also considered one of the natural numbers.

5. For example:
$$1 = 1$$
$$1 + 2 = 3$$
$$1 + 2 + 3 = 6$$
$$1 + 2 + 3 + 4 = 10$$
$$1 + 2 + 3 + 4 + 5 = 15$$
$$1 + 2 + 3 + 4 + 5 + 6 = 21$$

which are all triangular numbers, that is, numbers that can represent the number of dots that can be arranged in the form of an equilateral triangle. Triangular numbers will be investigated in greater detail beginning on page 42.

6. A prime number is a number (other than 0 or 1) that is divisible only by itself and 1. One can also say that a prime number is a number with exactly two (positive, integral) divisors.

The number 666 is equal to the sum of the cubes of the digits of its square, plus the digits of its cube. This means that if we find the square and the cube of 666:

$666^2 = 443,556$
$666^3 = 295,408,296$

and then take the sum of the cubes of the digits of the square of 666:

$4^3 + 4^3 + 3^3 + 5^3 + 5^3 + 6^3 = 621$

and add it to the sum of the digits of cube of 666:

$2 + 9 + 5 + 4 + 0 + 8 + 2 + 9 + 6 = 45$

we then get $621 + 45 = 666$.

Another peculiarity of 666 is that if we consider the prime factors of each of the two consecutive numbers 20,772,199 and 20,772,200, each of the sums of their prime factors is 666. That is:

$20,772,199 = 7 \cdot 41 \cdot 157 \cdot 461 \longrightarrow 7 + 41 + 157 + 461 = 666$
$20,772,200 = 2 \cdot 2 \cdot 2 \cdot 5 \cdot 5 \cdot 283 \cdot 367 \longrightarrow 2 + 2 + 2 + 5 + 5 + 283 + 367 = 666$

A strange occurrence of the number 666 is when we try to insert + signs into the sequence of numbers 1, 2, 3, 4, 5, 6, 7, 8, 9.

Here are two ways to do it:

$1 + 2 + 3 + 4 + 567 + 89 = 666$
or $123 + 456 + 78 + 9 = 666$

If we consider the reverse sequence 9, 8, 7, 6, 5, 4, 3, 2, 1, we can do it as follows:

$9 + 87 + 6 + 543 + 21 = 666$

The value of π is known to about 1.24 trillion places, where the decimal digits follow no discernable pattern.[7] Yet, strangely enough, when you take the sum of the first 144 decimal places,[8] you find it to be 666.

---

7. See A. S. Posamentier and I. Lehmann, π: *A Biography of the World's Most Mysterious Number* (Amherst, NY: Prometheus Books, 2004).
8. We refer here to the numerals after the decimal point.

$\pi \approx$ 3.14159265358979323846264338327950288419716939937510582097494 4592 307816406286208998628034825342117067982148086513282306647093 8446 09550582231725359

The sum of the digits is: $1 + 4 + 1 + 5 + 9 + 2 + 6 + 5 + 3 + 5 + 8 + 9 + 7 + 9 + 3 + 2 + 3 + 8 + 4 + 6 + 2 + 6 + 4 + 3 + 3 + 8 + 3 + 2 + 7 + 9 + 5 + 0 + 2 + 8 + 8 + 4 + 1 + 9 + 7 + 1 + 6 + 9 + 3 + 9 + 9 + 3 + 7 + 5 + 1 + 0 + 5 + 8 + 2 + 0 + 9 + 7 + 4 + 9 + 4 + 4 + 5 + 9 + 2 + 3 + 0 + 7 + 8 + 1 + 6 + 4 + 0 + 6 + 2 + 8 + 6 + 2 + 0 + 8 + 9 + 9 + 8 + 6 + 2 + 8 + 0 + 3 + 4 + 8 + 2 + 5 + 3 + 4 + 2 + 1 + 1 + 7 + 0 + 6 + 7 + 9 + 8 + 2 + 1 + 4 + 8 + 0 + 8 + 6 + 5 + 1 + 3 + 2 + 8 + 2 + 3 + 0 + 6 + 6 + 4 + 7 + 0 + 9 + 3 + 8 + 4 + 4 + 6 + 0 + 9 + 5 + 5 + 0 + 5 + 8 + 2 + 2 + 3 + 1 + 7 + 2 + 5 + 3 + 5 + 9 = 666$.

This remarkable—and sometimes "unlucky"—number seems to almost have a boundless array of number "coincidences" embedded within, such as that the sum of the numbers on a roulette wheel is 666.

Here are a few delectable number relationships that lead to 666:

$$666 = 1^6 - 2^6 + 3^6$$
$$666 = (6 + 6 + 6) + (6^3 + 6^3 + 6^3)$$
$$666 = (6^4 - 6^4 + 6^4) - (6^3 + 6^3 + 6^3) + (6 + 6 + 6)$$
$$666 = 5^3 + 6^3 + 7^3 - (6 + 6 + 6)$$
$$666 = 2^1 \cdot 3^2 + 2^3 \cdot 3^4$$

We can even generate 666 by representing each of its three digits in terms of 1, 2, and 3:

$$6 = 1 + 2 + 3$$
$$6 = 1 \cdot 2 \cdot 3$$
$$6 = \sqrt{1^3 + 2^3 + 3^3}$$

Therefore, $666 = (100)(1 + 2 + 3) + (10)(1 \cdot 2 \cdot 3) + \left(\sqrt{1^3 + 2^3 + 3^3} \right)$

The number 666 is also related to the Fibonacci numbers in a variety of ways. Consider the following, where F_n is the nth Fibonacci number:

$$F_1 - F_9 + F_{11} + F_{15} = 1 + 1 + 27 + 125 + 512 = 666$$

and when you inspect the subscripts, you get:

$$1 - 9 + 11 + 15 = 6 + 6 + 6$$

Similarly, for the cubes of the Fibonacci numbers:

$$F_1^3 + F_2^3 + F_4^3 + F_5^3 + F_6^3 = 1 + 1 + 27 + 125 + 512 = 666$$

and now the subscripts give us:

$$1 + 2 + 4 + 5 + 6 = 6 + 6 + 6$$

Our fascination with the number 666 is just to exhibit the beauty that lies in much of mathematics. Exploring the recreational side of mathematics is an enjoyable by-product of the important role it plays in all of scientific exploration and discovery.

PALINDROMIC NUMBERS

The number 666, as we mentioned earlier, is a palindrome. So we can use this number as a springboard to the next of our fascinating numerical amusements—where mathematics parallels amusing word games. A palindrome in mathematics is a number, such as 666 or 123,321, that reads the same in either direction. A palindrome can also be a word, phrase, or sentence that reads the same in both directions. Here are a few amusing palindromes:

EVE

RADAR

REVIVER

ROTATOR

LEPERS REPEL

MADAM I'M ADAM

STEP NOT ON PETS

DO GEESE SEE GOD

PULL UP IF I PULL UP

NO LEMONS, NO MELON

DENNIS AND EDNA SINNED

ABLE WAS I ERE I SAW ELBA

A MAN, A PLAN, A CANAL, PANAMA

A SANTA LIVED AS A DEVIL AT NASA

SUMS ARE NOT SET AS A TEST ON ERASMUS

ON A CLOVER, IF ALIVE, ERUPTS A VAST, PURE EVIL; A FIRE VOLCANO

There is a well-known Latin palindromic sentence that stems from the second century CE and has an additional amazing property. It reads: "Sator arepo tenet opera rotas," which commonly translates to "Arepo the sower holds the wheels at work." When the letters are placed in a five-by-five square arrangement (see figure 1-2), you can read the sentence in all directions. This is quite astonishing!

S	A	T	O	R
A	R	E	P	O
T	E	N	E	T
O	P	E	R	A
R	O	T	A	S

Figure 1-2

Palindromic numbers or numerical expressions can lead us to consider that dates can be a source for some symmetric inspection. For example, the year 2002 is a palindrome, as is 1991.[9] There were several dates in October 2001 that appeared as palindromes when written in the American style: 10/1/01, 10/22/01, and others. Europeans had the ultimate palindromic moment at 8:02 PM on February 20, 2002, since they would have written it as 20:02, 20.02.2002.

Looking further, the first four powers of 11 are palindromic numbers:

$11^0 = 1$

$11^1 = 11$

$11^2 = 121$

$11^3 = 1,331$

$11^4 = 14,641$

A palindromic number can either be a prime number or a composite number.[10] For example, 151 is a prime palindrome and 171 is a composite palindrome, since $171 = 3 \cdot 3 \cdot 19$. Yet, with the exception of 11, a palindromic prime must have an odd number of digits.

It is interesting to see how a palindromic number can be generated from other given numbers. All you need to do is to continually add a number to its reversal (i.e., the number written in the reverse order of digits) until you arrive at a palindrome.

For example, a palindrome can be reached with a single addition, such as with the starting number 23:

$23 + 32 = 55$, a palindrome

---

9. Those of us who have experienced 1991 and 2002 are part of the last generation who will have lived through two palindromic years for over the next one thousand years (assuming the current level of longevity).

10. The natural numbers are partitioned into three categories: the prime numbers—those that have only two factors, 1 and the number itself; the number 1; and all the rest are called composite numbers.

Or it might take two steps, such as with the starting number 75:
75 + 57 = 132 and 132 + 231 = 363, a palindrome

Or it might take three steps, such as with the starting number 86:
86 + 68 = 154, 154 + 451 = 605, and 605 + 506 = 1,111, a palindrome

The starting number 97 will require six steps to reach a palindrome, as you can see here:
97 + 79 = 176, 176 + 671 = 847, 847 + 748 = 1,595, 1,595 + 5,951 = 7,546, 7,546 + 6,457 = 14,003, and 14,003 + 30,041 = **44,044**

The number 98 will require twenty-four steps to reach a palindrome.

Be cautioned about using the starting number 196; this one has not yet been shown to produce a palindrome number—even with more than three million reversal additions. We still do not know if this one will ever reach a palindrome. If you were to try to apply this procedure on 196, you would eventually—at the sixteenth addition—reach the number 227,574,622, which you would also reach at the fifteenth step of the attempt to get a palindrome from the starting number 788. This would then tell you that applying the procedure to the number 788 has also never been shown to reach a palindrome. As a matter of fact, among the first 100,000 natural numbers, there are 5,996 numbers for which we have not yet been able to show that the procedure of reversal additions will lead to a palindrome. Some of these are 196; 691; 788; 887; 1,675; 5,761; 6,347; and 7,436.

We can arrive at some lovely patterns when dealing with palindromic numbers. For example, numbers that when cubed yield palindromic numbers are palindromic themselves.

DIVISIBILITY OF NUMBERS

In the base-10 number system, we are able to determine by inspection (and sometimes with a bit of simple arithmetic) when a given number is divisible by other numbers. For example, we know that when the last digit of a number is an even number, then the number is divisible by 2—such as with 30, 32, 34, 36, and 38. Of course, if the last digit is not divisible by 2, then we know that we cannot divide the number exactly by 2.

Divisibility by powers of 2

We can extend this to determining when a number is divisible by 4. When a number's last *two* digits (considered as a number) is divisible by 4, then, and only then, is the entire number also divisible by 4. For example, 1<u>24</u>, 1<u>28</u>, 3<u>56</u>, and 7<u>68</u> are each divisible by 4, while the number 322 is not divisible by 4, since 22 is not divisible by 4.

Furthermore, we can conclude that if, and only if, the last *three* digits of a number (considered as a number) is divisible by 8, the entire number is also divisible by 8. A clever person would then extend this rule to a number whose last *four* digits form a number that is divisible by 16 to conclude that then, and only then, is the entire number divisible by 16, and so on for succeeding powers of 2.

Divisibility by powers of 5

An analogous rule to that for powers of 2 can be used for divisibility by powers of 5; namely, that only if the last digit is either a 5 or a 0 is the number divisible by 5. Again only when the last two digits (considered as a number) is divisible by 25 is the number divisible by 25. Some such examples are 3<u>25</u>, 4<u>50</u>, 6<u>75</u>, and 8<u>00</u>. This rule continues for powers of 5 (5, 25, 125, 625, etc.) just as it did for powers of 2 earlier.

Divisibility by 3 and 9

For divisibility by 3, a different rule is used. Here we inspect the sum of the digits of the number to be considered. Only if the sum of the digits of the number being considered is divisible by 3 will the entire number be divisible by 3. For example, to determine if the number 345,678 is divisible by 3, we simply check to see if the sum of the digits $3 + 4 + 5 + 6 + 7 + 8 = 33$ is divisible by 3. In this case, it is; therefore, the number 345,678 is divisible by 3.

A similar rule can be used to determine divisibility by 9; namely, only if the sum of the digits of a given number is divisible by 9 is the number divisible by 9. An illustration of this is when we check the number 825,372 for divisibility by 9. The sum of the digits is $8 + 2 + 5 + 3 + 7 + 2 = 27$, which is divisible by 9. Therefore, the number 825,372 is divisible by 9.

You can see why these rules work when we consider the number 825,372 and represent it as:

$$825,372 = 8 \cdot (99,999 + 1) + 2 \cdot (9,999 + 1) + 5 \cdot (999 + 1) + 3 \cdot (99 + 1) + 7 \cdot (9 + 1) + 2$$
$$= (8 \cdot 99,999 + 2 \cdot 9,999 + 5 \cdot 999 + 3 \cdot 99 + 7 \cdot 9) + (\mathbf{8 + 2 + 5 + 3 + 7 + 2})$$

We can see that the term $(8 \cdot 99,999 + 2 \cdot 9,999 + 5 \cdot 999 + 3 \cdot 99 + 7 \cdot 9)$ is a multiple of 9 (and a multiple of 3 as well). Therefore, we only need to have the remainder of the number $(8 + 2 + 5 + 3 + 7 + 2)$ to be a multiple of 9 (or 3)—which just happens to be the sum of the digits of the original number 825,372—in order for the entire number to be divisible by 9 (or

3). In this case, $8 + 2 + 5 + 3 + 7 + 2 = 27$, which is divisible by 9 and 3. Hence the number 825,372 is divisible by 9 and 3, since the sum of its digits is also divisible by 9 and 3.

For example, the number 789 is not divisible by 9, because $7 + 8 + 9 = 24$; and 24 is not divisible by 9; yet the number 789 is divisible by 3, since 24 is divisible by 3.

Divisibility by composite numbers

We now have established divisibility rules to test for the numbers up to 10—with the exception of 6 and 7. But before we consider a test for divisibility by 7, we ought to make a statement about divisibility testing of composite (nonprime) numbers. To test divisibility by a composite number, we use the divisibility test for its "relatively prime factors"—that is, numbers whose only common factor is 1. For example, the test for divisibility by the composite number 12 would require applying the divisibility test for 3 and 4, which are its relatively prime factors (not 2 and 6, which are not relatively prime). The divisibility test for 18 requires applying the tests for divisibility by 2 and 9—which are relatively prime— and not the rules for divisibility by 3 and 6, whose product is also 18 but which are not relatively prime factors since they have a common factor of 3.

We can summarize the divisibility by composite numbers by inspecting the table below (figure 1-3), which shows the first few composite numbers and their relatively prime factors.

To be divisible by	6	10	12	14	15	18	20	21	24	26
The number must be divisible by	2, 3	2, 5	3, 4	2, 7	3, 5	2, 9	4, 5	3, 7	3, 8	2, 13

Figure 1-3

We now ought to consider divisibility by prime numbers.

Divisibility by other prime numbers

When we get to testing for divisibility by other prime numbers, we find that the rules may be a bit cumbersome and not realistic for use in everyday life situations—especially since the calculator is so pervasive. We will, therefore, present these divisibility rules largely for entertainment purposes, rather than as a useful tool.

Let us consider the rule for divisibility by 7 and then, as we inspect it, see how this can be generalized for other prime numbers.

The rule for divisibility by 7

Delete the last digit from the given number, and then subtract *twice* this deleted digit from the remaining number. If, and only if, the result is divisible by 7 will the original number be divisible by 7. This process may be repeated if the result is yet too large for simple visual inspection of divisibility by 7.

To best understand this divisibility test, we will apply it to determine if the number 876,547 is divisible by 7—without actually doing the division.

We begin with 876,547 and delete its units digit, 7, and then subtract its double, 14, from the remaining number:

$$87,654 - 14 = 87,640$$

Since we cannot yet determine if 87,640 is divisible by 7 through visual inspection, we shall continue the process.

We take this resulting number, 87,640, and delete its units digit, 0, and subtract its double, still 0, from the remaining number; we get: $8,764 - 0 = 8,764$.

This did not help us much in this case, so we shall continue the process. We delete the units digit, 4, from this resulting number, 8,764, and subtract its double, 8, from the remaining number to get: $876 - 8 = 868$. Since we still cannot visually inspect the resulting number, 868, for divisibility by 7, we continue the process.

This time we delete the units digit, 8, from the resulting number, 868, and subtract its double,16, from the remaining number to get: $86 - 16 = 70$, which we can easily determine is divisible by 7. Therefore, the original number, 876,547, is divisible by 7.

Now for the beauty of mathematics! That is, showing why this engaging procedure actually does what we say it does—that is, test for divisibility by 7. To be able to show why it works is the wonderful thing about mathematics.

To justify the technique of determining divisibility by 7, consider the various possible terminal digits (that you are "dropping") and the corresponding subtraction that is actually being done by dropping the last digit. In figure 1-4 you will see how dropping the terminal digit and doubling it to get the units digit of the number being subtracted gives us in each case a multiple of 7. That is, you can actually consider this as taking "bundles of 7" away from the original number. When a number is separated into parts and each part is divisible by 7, then the original number is divisible by 7. Therefore, if the remaining number (after the "bundles" of 7 have been removed) is divisible by 7, then so is the original number divisible by 7.

Terminal digit	Number subtracted from original
1	$20 + 1 = \ \ 21 = \ \ 3 \cdot 7$
2	$40 + 2 = \ \ 42 = \ \ 6 \cdot 7$
3	$60 + 3 = \ \ 63 = \ \ 9 \cdot 7$
4	$80 + 4 = \ \ 84 = 12 \cdot 7$
5	$100 + 5 = 105 = 15 \cdot 7$
6	$120 + 6 = 127 = 18 \cdot 7$
7	$140 + 7 = 147 = 21 \cdot 7$
8	$160 + 8 = 168 = 24 \cdot 7$
9	$180 + 9 = 189 = 27 \cdot 7$

Figure 1-4

The rule for divisibility by 11

The rule for divisibility by the next prime number, 11, can be determined in a similar manner as that for divisibility by 7. But in this case that technique becomes trivial and not particularly helpful. Still, since 11 is one more than the base (10), we have a rather different test for divisibility by 11. The procedure is to find the sums of the alternate digits and then take the difference of these two sums. If, and only if, that difference is divisible by 11, then the original number is divisible by 11. To better grasp this technique, we shall use an example. Suppose we would like to determine if the number 246,863,727 is divisible by 11. First, we find the sums of the alternate digits: $2 + 6 + 6 + 7 + 7 = 28$, and $4 + 8 + 3 + 2 = 17$. The difference of these two sums is $28 - 17 = 11$, which is clearly divisible by 11. Therefore, the original number 246,863,727 is divisible by 11.

To see how this rule for divisibility by 11 works, we shall represent the number 42,372,935 in the following way:

$$42,372,935 = 4 \cdot 10^7 + 2 \cdot 10^6 + 3 \cdot 10^5 + 7 \cdot 10^4 + 2 \cdot 10^3 + 9 \cdot 10^2 + 3 \cdot 10^1 + 5 \cdot 10^0$$

$$= 4 \cdot (10^7 + 1 - 1) + 2 \cdot (10^6 + 1 - 1) + 3 \cdot (10^5 + 1 - 1) + 7 \cdot (10^4 + 1 - 1)$$
$$+ 2 \cdot (10^3 + 1 - 1) + 9 \cdot (10^2 + 1 - 1) + 3 \cdot (10^1 + 1 - 1) + 5 \cdot 1$$

$$= 4 \cdot (10^7 + 1) - 4 \cdot 1 + 2 \cdot (10^6 - 1) + 2 \cdot 1 + 3 \cdot (10^5 + 1) - 3 \cdot 1 + 7 \cdot (10^4 - 1) + 7 \cdot 1$$
$$+ 2 \cdot (10^3 + 1) - 2 \cdot 1 + 9 \cdot (10^2 - 1) + 9 \cdot 1 + 3 \cdot (10^1 + 1) - 3 \cdot 1 + 5 \cdot 1$$

$$= 4 \cdot (\mathbf{10^7 + 1}) - 4 \cdot 1 + 2 \cdot (\mathbf{10^6 - 1}) + 2 \cdot 1 + 3 \cdot (\mathbf{10^5 + 1}) - 3 \cdot 1 + 7 \cdot (\mathbf{10^4 - 1}) + 7 \cdot 1$$
$$+ 2 \cdot (\mathbf{10^3 + 1}) - 2 \cdot 1 + 9 \cdot (\mathbf{10^2 - 1}) + 9 \cdot 1 + 3 \cdot (\mathbf{10^1 + 1}) - 3 \cdot 1 + 5 \cdot 1$$

Each of the bold terms, $(10^7 + 1)$, $(10^6 - 1)$, $(10^5 + 1)$, $(10^4 - 1)$, $(10^3 + 1)$, $(10^2 - 1)$, $(10^1 + 1)$, is divisible by 11. Therefore, we need to just insure that the sum of the rest of the terms is also divisible by 11. They are: $-4\cdot1 + 2\cdot1 - 3\cdot1 + 7\cdot1 - 2\cdot1 + 9\cdot1 - 3\cdot1 + 5\cdot1$ $= -4 + 2 - 3 + 7 - 2 + 9 - 3 + 5 = 11$. This is the difference of the sums of the alternate digits, $(2 + 7 + 9 + 5) - (4 + 3 + 2 + 3)$, of the number 42,372,935.

We will use the procedure for divisibility by 7 as a guide to discover an analogous rule for divisibility by the prime number 13. This is similar to the rule for testing divisibility by 7, except that the 7 is replaced by 13 and instead of subtracting twice the deleted digit, we subtract nine times the deleted digit each time.

The rule for divisibility by 13

Delete the last digit from the given number, and then subtract *nine times* this deleted digit from the remaining number. If, and only if, the result is divisible by 13 will the original number be divisible by 13. This process may be repeated if the result is too large for simple inspection of divisibility by 13.

Let's check the number 5,616 for divisibility by 13. Begin with 5,616 and delete its units digit, 6, and subtract nine times it, 54, from the remaining number: $561 - 54 = 507$.

Since we still cannot visually inspect the resulting number for divisibility by 13, we continue the process.

Continue with the resulting number, 507, and delete its units digit and subtract nine times this digit from the remaining number: $50 - 63 = -13$, which is divisible by 13. Therefore, the original number is divisible by 13.

To determine the "multiplier," 9, we sought the smallest multiple of 13 that ends in a 1. That was 91, where the tens digit is 9 times the units digit. Once again, consider the various possible terminal digits and the corresponding subtractions in the following table (figure 1-5).

Terminal digit	Number subtracted from original
1	$90 + 1 = 91 = 7\cdot13$
2	$180 + 2 = 182 = 14\cdot13$
3	$270 + 3 = 273 = 21\cdot13$
4	$360 + 4 = 364 = 28\cdot13$
5	$450 + 5 = 455 = 35\cdot13$
6	$540 + 6 = 546 = 42\cdot13$
7	$630 + 7 = 637 = 49\cdot13$
8	$720 + 8 = 728 = 56\cdot13$
9	$810 + 9 = 819 = 63\cdot13$

Figure 1-5

In each case, a multiple of 13 is being subtracted one or more times from the original number. Hence, only if the remaining number is divisible by 13 is the original number divisible by 13.

As we proceed to the divisibility test for the next prime, 17, we shall once again use this technique. We seek the multiple of 17 with a units digit of 1, that is, 51. This gives us the "multiplier," 5, we need to establish the following rule.

The rule for divisibility by 17

Delete the units digit and subtract *five times* the deleted digit each time from the remaining number until you reach a number small enough to determine if it is divisible by 17.

We can justify the rule for divisibility by 17 as we did the rules for 7 and 13. Each step of the procedure subtracts a "bunch of 17s" from the original number until we reduce the number to a manageable size and can make a visual inspection of divisibility by 17.

The patterns developed in the preceding three divisibility rules (for 7, 13, and 17) should lead you to develop similar rules for testing divisibility by larger primes. The following chart (figure 1-6) presents the "multipliers" of the deleted digits for various primes.

To test divisibility by	7	11	13	17	19	23	29	31	37	41	43	47
Multiplier	2	1	9	5	17	16	26	3	11	4	30	14

Figure 1-6

You may want to extend this chart; it's fun, and it will enhance your perception of mathematics. In addition to extending the rules for divisibility by prime numbers, you may also want to extend your knowledge of divisibility rules to include composite (i.e., nonprime) numbers. Remember that to test divisibility for composite numbers, we need to consider the rules for the numbers' relatively prime factors—this guarantees that we will be using independent divisibility rules.

NUMBERS AND THEIR DIGITS

Our number system has many unusual features, which are often well hidden. Discovering them can certainly be a rewarding and entertaining experience. Sometimes we just stumble onto these relationships, while at other times, the discovery of these relationships is the result of experimentation and searching—often based on a hunch.

Consider the following relationship and describe what is going on here:

$$81 = (8 + 1)^2 = 9^2$$

We have taken *the square of the sum of the digits.*

Here is another such example:

$$4,913 = (4 + 9 + 1 + 3)^3 = 17^3$$

In both cases, we have taken the sum of the digits to a power and ended up with the number we started with. Impressed? You ought to be, for this is quite astonishing. Trying to find other such number relationships is no mean feat. For example, $82 \neq (8 + 2)^2 = 10^2 = 100$ does not follow this pattern. The list below (figure 1-7) will provide you with lots of examples of these unusual numbers. The beauty is self-evident.

Number	=	(Sum of the digits)^n		
81	=	$(8+1)^2$	=	9^2
512	=	$(5+1+2)^3$	=	8^3
4,913	=	$(4+9+1+3)^3$	=	17^3
5,832	=	$(5+8+3+2)^3$	=	18^3
17,576	=	$(1+7+5+7+6)^3$	=	26^3
19,683	=	$(1+9+6+8+3)^3$	=	27^3
2,401	=	$(2+4+0+1)^4$	=	7^4
234,256	=	$(2+3+4+2+5+6)^4$	=	22^4
390,625	=	$(3+9+0+6+2+5)^4$	=	25^4
614,656	=	$(6+1+4+6+5+6)^4$	=	28^4
1,679,616	=	$(1+6+7+9+6+1+6)^4$	=	36^4
17,210,368	=	$(1+7+2+1+0+3+6+8)^5$	=	28^5
52,521,875	=	$(5+2+5+2+1+8+7+5)^5$	=	35^5
60,466,176	=	$(6+0+4+6+6+1+7+6)^5$	=	36^5
205,962,976	=	$(2+0+5+9+6+2+9+7+6)^5$	=	46^5
34,012,224	=	$(3+4+0+1+2+2+2+4)^6$	=	18^6
8,303,765,625	=	$(8+3+0+3+7+6+5+6+2+5)^6$	=	45^6
24,794,911,296	=	$(2+4+7+9+4+9+1+1+2+9+6)^6$	=	54^6
68,719,476,736	=	$(6+8+7+1+9+4+7+6+7+3+6)^6$	=	64^6

Number	=	(Sum of the digits)^n		
612,220,032	=	$(6+1+2+2+2+0+0+3+2)^7$	=	18^7
10,460,353,203	=	$(1+0+4+6+0+3+5+3+2+0+3)^7$	=	27^7
27,512,614,111	=	$(2+7+5+1+2+6+1+4+1+1+1)^7$	=	31^7
52,523,350,144	=	$(5+2+5+2+3+3+5+0+1+4+4)^7$	=	34^7
271,818,611,107	=	$(2+7+1+8+1+8+6+1+1+1+0+7)^7$	=	43^7
1,174,711,139,837	=	$(1+1+7+4+7+1+1+1+3+9+8+3+7)^7$	=	53^7
2,207,984,167,552	=	$(2+2+0+7+9+8+4+1+6+7+5+5+2)^7$	=	58^7
6,722,988,818,432	=	$(6+7+2+2+9+8+8+8+1+8+4+3+2)^7$	=	68^7
20,047,612,231,936	=	$(2+0+0+4+7+6+1+2+2+3+1+9+3+6)^8$	=	46^8
72,301,961,339,136	=	$(7+2+3+0+1+9+6+1+3+3+9+1+3+6)^8$	=	54^8
248,155,780,267,521	=	$(2+4+8+1+5+5+7+8+0+2+6+7+5+2+1)^8$	=	63^8

Figure 1-7

And here is another one:

$$20{,}864{,}448{,}472{,}975{,}628{,}947{,}226{,}005{,}981{,}267{,}194{,}447{,}042{,}584{,}001 =$$
$$(2+0+8+6+4+4+4+8+4+7+2+9+7+5+6+2+8+9+4+7+2+2$$
$$+6+0+0+5+9+8+1+2+6+7+1+9+4+4+4+7+0+4+2+5+8+4$$
$$+0+0+1)^{20} = 207^{20}$$

BEAUTIFUL NUMBER RELATIONSHIPS

There are times when the numbers speak more effectively for themselves than any explanation. Here is one such case. Just look at these equalities and see if you can discover others of the same type.

$$1^1 + 6^1 + 8^1 = 15 = 2^1 + 4^1 + 9^1$$
$$1^2 + 6^2 + 8^2 = 101 = 2^2 + 4^2 + 9^2$$

$$1^1 + 5^1 + 8^1 + 12^1 = 26 = 2^1 + 3^1 + 10^1 + 11^1$$
$$1^2 + 5^2 + 8^2 + 12^2 = 234 = 2^2 + 3^2 + 10^2 + 11^2$$
$$1^3 + 5^3 + 8^3 + 12^3 = 2{,}366 = 2^3 + 3^3 + 10^3 + 11^3$$

$$1^1 + 5^1 + 8^1 + 12^1 + 18^1 + 19^1 = 63 = 2^1 + 3^1 + 9^1 + 13^1 + 16^1 + 20^1$$
$$1^2 + 5^2 + 8^2 + 12^2 + 18^2 + 19^2 = 919 = 2^2 + 3^2 + 9^2 + 13^2 + 16^2 + 20^2$$
$$1^3 + 5^3 + 8^3 + 12^3 + 18^3 + 19^3 = 15{,}057 = 2^3 + 3^3 + 9^3 + 13^3 + 16^3 + 20^3$$
$$1^4 + 5^4 + 8^4 + 12^4 + 18^4 + 19^4 = 260{,}755 = 2^4 + 3^4 + 9^4 + 13^4 + 16^4 + 20^4$$

ARMSTRONG NUMBERS

Some numbers, often referred to as Armstrong numbers,[11] are each equal to the sum of their digits, each of which is taken to the power equal to the number of digits in the original number.

For example, the three-digit number $153 = 1^3 + 5^3 + 3^3$. Strangely enough, while there are no two-digit Armstrong numbers, there are four three-digit Armstrong numbers as follows (figure 1-8):

n	$100 \leq n \leq 999$	
153	$1^3 + 5^3 + 3^3 = 1 + 125 + 27$	$= 153$
370	$3^3 + 7^3 + 0^3 = 27 + 343 + 0$	$= 370$
371	$3^3 + 7^3 + 1^3 = 27 + 343 + 1$	$= 371$
407	$4^3 + 0^3 + 7^3 = 64 + 0 + 343$	$= 407$

Figure 1-8

There are three four-digit Armstrong numbers as seen below (figure 1-9).

n	$1,000 \leq n \leq 9,999$	
1,634	$1^4 + 6^4 + 3^4 + 4^4 = 1 + 1,296 + 81 + 256$	$= 1,634$
8,208	$8^4 + 2^4 + 0^4 + 8^4 = 4,096 + 16 + 0 + 4,096$	$= 8,208$
9,474	$9^4 + 4^4 + 7^4 + 4^4 = 6,561 + 256 + 2,401 + 256$	$= 9,474$

Figure 1-9

Among the five-digit numbers, there are three Armstrong numbers (figure 1-10).

n	$10,000 \leq n \leq 99,999$	
54,748	$5^5 + 4^5 + 7^5 + 4^5 + 8^5 = 3,125 + 1,024 + 16,807 + 1,024 + 32,768$	$= 54,748$
92,727	$9^5 + 2^5 + 7^5 + 2^5 + 7^5 = 59,049 + 32 + 16,807 + 32 + 16,807$	$= 92,727$
93,084	$9^5 + 3^5 + 0^5 + 8^5 + 4^5 = 59,049 + 243 + 0 + 32,768 + 1,024$	$= 93,084$

Figure 1-10

The only Armstrong number among the six-digit numbers is: 548,834.

$5^6 + 4^6 + 8^6 + 8^6 + 3^6 + 4^6 = 15,625 + 4,096 + 262,144 + 262,144 + 729 + 4,096 = 548,834$

---

11. L. Deimel Jr. and M. Jones, "Finding Pluperfect Digital Invariants," *Journal of Recreational Mathematics* 42, no. 2 (1981–82): 87–107.

The following is a table (figure 1-11) that lists the Armstrong numbers of 7–10 digits.

k	Armstrong numbers
7	$n =$ 1,741,725; 4,210,818; 9,800,817; 9,926,315
8	$n =$ 24,678,050; 24,678,051; 88,593,477
9	$n =$ 146,511,208; 472,335,975; 534,494,836; 912,985,153
10	$n =$ 4,679,307,774

Figure 1-11

Here is a number that is 39 digits long:

115,132,219,018,763,992,565,095,597,973,971,522,401

and is equal to the sum of its digits, each of which is taken to the thirty-ninth power, thus making it an Armstrong number. Furthermore, it is the largest Armstrong number. In other words,

$$1^{39} + 1^{39} + 5^{39} + 1^{39} + 3^{39} + 2^{39} + 2^{39} + 1^{39} + 9^{39} + 0^{39} + 1^{39} + 8^{39} + 7^{39} + 6^{39} + 3^{39} + 9^{39}$$
$$+ 9^{39} + 2^{39} + 5^{39} + 6^{39} + 5^{39} + 0^{39} + 9^{39} + 5^{39} + 5^{39} + 9^{39} + 7^{39} + 9^{39} + 7^{39} + 3^{39} + 9^{39}$$
$$+ 7^{39} + 1^{39} + 5^{39} + 2^{39} + 2^{39} + 4^{39} + 0^{39} + 1^{39}$$
$$= 115,132,219,018,763,992,565,095,597,973,971,522,401$$
$$\approx 1.151322190 \cdot 10^{38}$$

The Armstrong Numbers

k	Armstrong numbers of k-digit lengths
1	0; 1; 2; 3; 4; 5; 6; 7; 8; 9
3	153; 370; 371; 407
4	1,634; 8,208; 9,474
5	54,748; 92,727; 93,084
6	548,834
7	1,741,725; 4,210,818; 9,800,817; 9,926,315
8	24,678,050; 24,678,051; 88,593,477
9	146,511,208; 472,335,975; 534,494,836; 912,985,153
10	4,679,307,774
11	32,164,049,650; 32,164,049,651; 40,028,394,225; 42,678,290,603; 44,708,635,679; 49,388,550,606; 82,693,916,578; 94,204,591,914
14	28,116,440,335,967

k	Armstrong numbers of k-digit lengths
16	4,338,281,769,391,370; 4,338,281,769,391,371
17	21,897,142,587,612,075; 35,641,594,208,964,132; 35,875,699,062,250,035
19	1,517,841,543,307,505,039; 3,289,582,984,443,187,032; 4,498,128,791,164,624,869; 4,929,273,885,928,088,826
20	63,105,425,988,599,693,916
21	128,468,643,043,731,391,252; 449,177,399,146,038,697,307
23	21,887,696,841,122,916,288,858; 27,879,694,893,054,074,471,405; 27,907,865,009,977,052,567,814; 28,361,281,321,319,229,463,398; 35,452,590,104,031,691,935,943
24	174,088,005,938,065,293,023,722; 188,451,485,447,897,896,036,875; 239,313,664,430,041,569,350,093
25	1,550,475,334,214,501,539,088,894; 1,553,242,162,893,771,850,669,378; 3,706,907,995,955,475,988,644,380; 3,706,907,995,955,475,988,644,381; 4,422,095,118,095,899,619,457,938
27	121,204,998,563,613,372,405,438,066; 121,270,696,006,801,314,328,439,376; 128,851,796,696,487,777,842,012,787; 174,650,464,499,531,377,631,639,254; 177,265,453,171,792,792,366,489,765
29	14,607,640,612,971,980,372,614,873,089; 19,008,174,136,254,279,995,012,734,740; 19,008,174,136,254,279,995,012,734,741; 23,866,716,435,523,975,980,390,369,295
31	1,145,037,275,765,491,025,924,292,050,346; 1,927,890,457,142,960,697,580,636,236,639; 2,309,092,682,616,190,307,509,695,338,915
32	17,333,509,997,782,249,308,725,103,962,772
33	186,709,961,001,538,790,100,634,132,976,990; 186,709,961,001,538,790,100,634,132,976,991
34	1,122,763,285,329,372,541,592,822,900,204,593
35	12,639,369,517,103,790,328,947,807,201,478,392; 12,679,937,780,272,278,566,303,885,594,196,922
37	1,219,167,219,625,434,121,569,735,803,609,966,019
38	12,815,792,078,366,059,955,099,770,545,296,129,367
39	115,132,219,018,763,992,565,095,597,973,971,522,400; 115,132,219,018,763,992,565,095,597,973,971,522,401

Figure 1-12

There are no Armstrong numbers for $k = 2, 12, 13, 15, 18, 22, 26, 28, 30,$ and 36 (and $k >$ 39). In fact, there are only eighty-nine Armstrong numbers in the decimal system.

The following is a list of the *consecutive* Armstrong numbers:

$k = 3$:   370; 371

$k = 8$:   24,678,050; 24,678,051

$k = 11$:   32,164,049,650; 32,164,049,651

$k = 16$:   4,338,281,769,391,370; 4,338,281,769,391,371

$k = 25$:   3,706,907,995,955,475,988,644,380; 3,706,907,995,955,475,988,644,381

$k = 29$:   19,008,174,136,254,279,995,012,734,740;

19,008,174,136,254,279,995,012, 734,741

$k = 33$:   186,709,961,001,538,790,100,634,132,976,990;

186,709,961,001,538,790,100, 634,132,976,991

$k = 39$:   115,132,219,018,763,992,565,095,597,973,971,522,400

115,132,219,018,763, 992,565,095,597,973,971,522,401

Incidentally, our first Armstrong number, 153, has some other amazing properties. It is a triangular number, where

$$1 + 2 + 3 + 4 + 5 + 6 + 7 + 8 + 9 + 10 + 11 + 12 + 13 + 14 + 15 + 16 + 17 = 153$$

The number 153 is not only equal to the sum of the cubes of its digits ($1^3 + 5^3 + 3^3 = 1 + 125 + 27 = 153$) but it is also a number that can be expressed as the sum of consecutive factorials:[12] $1! + 2! + 3! + 4! + 5! = 153$.

Can you discover other properties of this ubiquitous number 153?

MORE NUMBER CURIOSITIES

As an "extension" of the Armstrong numbers, we will present numbers where the sum of the digits of the original number—each taken to the power one more or less than the number of digits in the original number—is equal to the original number. This differs from the Armstrong numbers, where the power to which each of the digits was taken was equal to the number of digits in the original number.

To begin, consider the following four-digit numbers where each of the digits is take to the fifth power and the sum is equal to the original number.

$4,150 = 4^5 + 1^5 + 5^5 + 0^5$

$4,151 = 4^5 + 1^5 + 5^5 + 1^5$

---

12. $n!$ reads "n factorial" and equals $n \cdot (n - 1) \cdot (n - 2) \cdot \ldots \cdot 3 \cdot 2 \cdot 1$.

The following are some other examples of non-Armstrong numbers:

$194,979 = 1^5 + 9^5 + 4^5 + 9^5 + 7^5 + 9^5$
$14,459,929 = 1^7 + 4^7 + 4^7 + 5^7 + 9^7 + 9^7 + 2^7 + 9^7$

Here is a forty-one-digit number[13] that is equal to the sum of the digits, each taken to the forty-second power: 36,428,594,490,313,158,783,584,452,532,870,892,261,556

$$3^{42} + 6^{42} + 4^{42} + 2^{42} + 8^{42} + 5^{42} + 9^{42} + 4^{42} + 4^{42} + 9^{42} + 0^{42} + 3^{42} + 1^{42} + 3^{42} + 1^{42} + 5^{42}$$
$$+ 8^{42} + 7^{42} + 8^{42} + 3^{42} + 5^{42} + 8^{42} + 4^{42} + 4^{42} + 5^{42} + 2^{42} + 5^{42} + 3^{42} + 2^{42} + 8^{42} + 7^{42}$$
$$+ 0^{42} + 8^{42} + 9^{42} + 2^{42} + 2^{42} + 6^{42} + 1^{42} + 5^{42} + 5^{42} + 6^{42}$$
$$= 36,428,594,490,313,158,783,584,452,532,870,892,261,556$$

The relationships between numbers and powers can be fascinating. Here are a few to relish.

Notice how, in these cases, quite the opposite of the previous numbers, the powers reflect the original number and the base stays the same.

$4,624 = 4^4 + 4^6 + 4^2 + 4^4$
$1,033 = 8^1 + 8^0 + 8^3 + 8^3$
$595,968 = 4^5 + 4^9 + 4^5 + 4^9 + 4^6 + 4^8$
$3,909,511 = 5^3 + 5^9 + 5^0 + 5^9 + 5^5 + 5^1 + 5^1$
$13,177,388 = 7^1 + 7^3 + 7^1 + 7^7 + 7^7 + 7^3 + 7^8 + 7^8$
$52,135,640 = 19^5 + 19^2 + 19^1 + 19^3 + 19^5 + 19^6 + 19^4 + 19^0$

Then there are those numbers that differ from the previous examples and lend themselves to an even more astonishing pattern: here the powers and the bases match the original digits.[14]

$3,435 = 3^3 + 4^4 + 3^3 + 5^5$
$438,579,088 = 4^4 + 3^3 + 8^8 + 5^5 + 7^7 + 9^9 + 0^0 + 8^8 + 8^8$

We can also find a reversal of this pattern, namely, where the bases go in the proper order and the exponents progress in the reverse order.

$48,625 = 4^5 + 8^2 + 6^6 + 2^8 + 5^4$
$397,612 = 3^2 + 9^1 + 7^6 + 6^7 + 1^9 + 2^3$

---

13. See L. E. Deimel Jr. and M. T. Jones, *Journal of Recreational Mathematics* 14, no. 4 (1981–82): 284.
14. Although 0^0 is undefined, we shall value it at 0 for our purposes here.

Searching through our number system reveals that there are also some entertaining patterns where consecutive exponents are used:

$$43 = 4^2 + 3^3$$
$$63 = 6^2 + 3^3$$
$$89 = 8^1 + 9^2$$
$$1,676 = 1^5 + 6^4 + 7^3 + 6^2$$

Yet when we formalize this a bit and use the consecutive natural numbers as exponents, we have the following amazing relationships:

$$135 = 1^1 + 3^2 + 5^3$$
$$175 = 1^1 + 7^2 + 5^3$$
$$518 = 5^1 + 1^2 + 8^3$$
$$598 = 5^1 + 9^2 + 8^3$$
$$1,306 = 1^1 + 3^2 + 0^3 + 6^4$$
$$1,676 = 1^1 + 6^2 + 7^3 + 6^4$$
$$2,427 = 2^1 + 4^2 + 2^3 + 7^4$$
$$2,646,798 = 2^1 + 6^2 + 4^3 + 6^4 + 7^5 + 9^6 + 8^7$$

You may have noticed that 1,676 appeared in both lists of consecutive exponents.

A further charming occurrence in mathematics—in the decimal system, of course—is seen with the following illustrations. Although you may find this a bit contrived, it simply highlights the notion that there are probably limitless nuggets we can cull from our numbers system.

$$\mathbf{16}^3 + \mathbf{50}^3 + \mathbf{33}^3 = 4,096 + 125,000 + 35,937 = \mathbf{16\ 5{,}0\ 33} = 165,033$$
$$\mathbf{166}^3 + \mathbf{500}^3 + \mathbf{333}^3 = 4,574,296 + 125,000,000 + 36,926,037 = \mathbf{166,\ 500,\ 333}$$
$$= 166,500,333$$
$$\mathbf{588}^2 + \mathbf{2,353}^2 = 345,744 + 5,536,609 = \mathbf{5{,}88\ 2{,}353} = 5,882,353$$

MORE UNUSUAL NUMBER RELATIONSHIPS

There are a number of unusual relationships between certain numbers—as represented in the decimal system. There is not much explanation for them. So just enjoy them and see if you can find others like them.

We are going to present pairs of numbers where the product and the sum are reversals of each other (figure 1-13).

The two numbers		their *product*	their *sum*
9	9	81	18
3	24	72	27
2	47	94	49
2	497	994	499

Figure 1-13

Moving along to other number curiosities, consider the number 3,025. Suppose we split the number as 30 and 25, and then add these two numbers to get $30 + 25 = 55$. Then, if we square this result, we surprisingly get back to the original number: $55^2 = 3,025$. The question that immediately arises is "Are there other numbers for which this little procedure also works?"

Well, here are a few more such numbers:

$9,801 \rightarrow 98 + 01 = 99$	and	$99^2 = 9,801$
$2,025 \rightarrow 20 + 25 = 45$	and	$45^2 = 2,025$
$088,209 \rightarrow 088 + 209 = 297$	and	$297^2 = 88,209$
$494,209 \rightarrow 494 + 209 = 703$	and	$703^2 = 494,209$
$998,001 \rightarrow 998 + 001 = 999$	and	$999^2 = 998,001$
$99,980,001 \rightarrow 9998 + 0001 = 9,999$	and	$9,999^2 = 99,980,001$

Perhaps you can now find other numbers where this amazing relationship will hold.

While we are in the mode of "splitting" numbers, let's partition numbers to get some other fascinating relationships

$1,233 \rightarrow 12|\, 33 = 12^2 + 33^2$

$8,833 \rightarrow 88|\, 33 = 88^2 + 33^2$

$990,100 \rightarrow 990|\, 100 = 990^2 + 100^2$

$94,122,353 \rightarrow 9412|\, 2353 = 9,412^2 + 2,353^2$

$7,416,043,776 \rightarrow 74160|\, 43776 = 74,160^2 + 43,776^2$

$116,788,321,168 \rightarrow 116788|\, 321168 = 116,788^2 + 321,168^2$

$221,859 \rightarrow 22|\, 18|\, 59 = 22^3 + 18^3 + 59^3$

$166,500,333 \rightarrow 166|\, 500|\, 333 = 166^3 + 500^3 + 333^3$

You might like to verify some of the following that can be split, where the sum of the squares of the parts will equal the original number:

$$10,100 \rightarrow 10\,|\,100 = 10^2 + 100^2$$
$$5,882,353 \rightarrow 588\,|\,2353 = 588^2 + 2,353^2$$

Here are a few more such numbers: $9,901,0\,|\,09,901$, $1,765,0\,|\,38,125$, $2,584,0\,|\,43,776$, $999,900\,|\,010,000$, $123,288\,|\,328,768$, and there are still more. You might want to verify this relationship. You can also find more of these numbers.

An analogous situation—this time using differences instead of sums—also produces some interesting relationships:

$$48 \rightarrow 4\,|\,8 = 8^2 - 4^2$$
$$3,468 \rightarrow 34\,|\,68 = 68^2 - 34^2$$
$$416,768 \rightarrow 416\,|\,768 = 768^2 - 416^2$$
$$33,346,668 \rightarrow 3334\,|\,6668 = 6,668^2 - 3,334^2$$

You might like to verify some of the following that can be split, where the difference of the squares of the parts will equal the original number. $16\,|\,128$, $34\,|\,188$, $140\,|\,400$, $484\,|\,848$, $530\,|\,901$, $334\,|\,668$, $234\,|\,1,548$, and others.

AN UNUSUAL SITUATION

Sometimes numbers can be changed by accident. In a recent publication, a computer glitch dropped all the exponents to the base line. So that 8^3 was shown as 83. Would there be a value that would not be affected by such a computer failure? You can try all the possibilities and you may stumble upon $2^5 \cdot 9^2 = 2,592$, which is believed to be the only such case.

FRIENDLY NUMBERS

What could possibly make two numbers friendly? Your first reaction might be that this means numbers that are friendly to you. Actually, we will be inspecting numbers that are "friendly" to each other. Mathematicians have decided that two numbers are considered "friendly" (or as often used in the more sophisticated literature, "amicable") if the sum of the proper divisors[15] of one number equals the second number *and* the sum of the proper divisors of the second number equals the first number.

---

15. *Proper divisors* of a number n are divisors that are smaller than the number n. That means 1 is a proper divisor, but n is not.

Sounds complicated? It's not at all difficult to understand. Let's look at the smallest pair of friendly numbers: 220 and 284.

The divisors of **220** (other than 220 itself) are 1, 2, 4, 5, 10, 11, 20, 22, 44, 55, and 110. Their sum is $1 + 2 + 4 + 5 + 10 + 11 + 20 + 22 + 44 + 55 + 110 = $ **284**.

The divisors of **284** (other than 284 itself) are 1, 2, 4, 71, and 142, and their sum is $1 + 2 + 4 + 71 + 142 = $ **220**. This shows that the two numbers are friendly numbers.

Although this pair of friendly numbers was already known to Pythagoras about 500 BCE, a second pair of friendly numbers, first discovered in 1636 by the French mathematician Pierre de Fermat (1607–1665),[16] is 17,296 and 18,416.

$17,296 = 2^4 \cdot 23 \cdot 47$, and $18,416 = 2^4 \cdot 1,151$

The sum of the factors of 17,296 is
$1 + 2 + 4 + 8 + 16 + 23 + 46 + 47 + 92 + 94 + 184 + 188 + 368 + 376 + 752 + 1,081 + 2,162 + 4,324 + 8,648 = \underline{18,416}$

The sum of the factors of 18,416 is
$1 + 2 + 4 + 8 + 16 + 1,151 + 2,302 + 4,604 + 9,208 = \underline{17,296}$

Thus, they, too, are truly friendly numbers!

Two years later, the French mathematician René Descartes (1596–1650) found another pair of friendly numbers: 9,363,584 and 9,437,056. By 1747, the Swiss mathematician Leonhard Euler (1707–1783) discovered sixty pairs of friendly numbers, yet he seemed to have overlooked the second smallest pair, which was discovered in 1866 by the sixteen-year-old Nicolò Paganini:

The sum of the factors of 1,184 is $1 + 2 + 4 + 8 + 16 + 32 + 37 + 74 + 148 + 296 + 592 = 1,210$

And the sum of the factors of 1,210 is $1 + 2 + 5 + 10 + 11 + 22 + 55 + 110 + 121 + 242 + 605 = 1,184$

We now have identified over 363,000 pairs of friendly numbers. Here are a few of these:

2,620 and 2,924

5,020 and 5,564

6,232 and 6,368

10,744 and 10,856

12,285 and 14,595

63,020 and 76,084

111,448,537,712 and 118,853,793,424

---

16. There is evidence that Fermat's discovery was anticipated by the Morroccan mathematician Ibn al-Banna al-Marakushi al-Azdi (1256–ca. 1321).

You might want to verify the above pairs' "friendliness"!

For the experts, the following is one method for finding friendly numbers:

$$\text{Let} \quad a = 3 \cdot 2^n - 1$$
$$b = 3 \cdot 2^{n-1} - 1$$
$$c = 3^2 \cdot 2^{2n-1} - 1$$

where n is an integer ≥ 2 and where a, b, and c are all prime numbers, then $2^n \cdot ab$ and $2^n \cdot c$ are friendly numbers.

(Notice that for $n = 2, 4,$ and 7 that a, b, and c are all prime for $n \leq 200$.)

Here is a list of friendly numbers less than 10,000,000:

	First number	Second number	Year of discovery	Discoverer
1	220	284	-	Pythagoras
2	1,184	1,210	1860	Paganini
3	2,620	2,924	1747	Euler
4	5,020	5,564	1747	Euler
5	6,232	6,368	1747	Euler
6	10,744	10,856	1747	Euler
7	12,285	14,595	1939	Brown
8	17,296	18,416	ca. 1310/1636	Ibn al-Banna/Fermat
9	63,020	76,084	1747	Euler
10	66,928	66,992	1747	Euler
11	67,095	71,145	1747	Euler
12	69,615	87,633	1747	Euler
13	79,750	88,730	1964	Rolf
14	100,485	124,155	1747	Euler
15	122,265	139,815	1747	Euler
16	122,368	123,152	1941/42	Poulet
17	141,664	153,176	1747	Euler
18	142,310	168,730	1747	Euler
19	171,856	176,336	1747	Euler
20	176,272	180,848	1747	Euler
21	185,368	203,432	1966	Alanen/Ore/Stempel
22	196,724	202,444	1747	Euler
23	280,540	365,084	1966	Alanen/Ore/Stempel
24	308,620	389,924	1747	Euler
25	319,550	430,402	1966	Alanen/Ore/Stempel
26	356,408	399,592	1921	Mason
27	437,456	455,344	1747	Euler

	First number	Second number	Year of discovery	Discoverer
28	469,028	486,178	1966	Alanen/Ore/Stempel
29	503,056	514,736	1747	Euler
30	522,405	525,915	1747	Euler
31	600,392	669,688	1921	Mason
32	609,928	686,072	1747	Euler
33	624,184	691,256	1921	Mason
34	635,624	712,216	1921	Mason
35	643,336	652,664	1747	Euler
36	667,964	783,556	1966	Alanen/Ore/Stempel
37	726,104	796,696	1921	Mason
38	802,725	863,835	1966	Alanen/Ore/Stempel
39	879,712	901,424	1966	Alanen/Ore/Stempel
40	898,216	980,984	1747	Euler
41	947,835	1,125,765	1946	Escott
42	998,104	1,043,096	1966	Alanen/Ore/Stempel
43	1,077,890	1,099,390	1966	Lee
44	1,154,450	1,189,150	1957	Garcia
45	1,156,870	1,292,570	1946	Escott
46	1,175,265	1,438,983	1747	Euler
47	1,185,376	1,286,744	1929	Gerardin
48	1,280,565	1,340,235	1747	Euler
49	1,328,470	1,483,850	1966	Lee
50	1,358,595	1,486,845	1747	Euler
51	1,392,368	1,464,592	1747	Euler
52	1,466,150	1,747,930	1966	Lee
53	1,468,324	1,749,212	1967	Bratley/McKay
54	1,511,930	1,598,470	1946	Escott
55	1,669,910	2,062,570	1966	Lee
56	1,798,875	1,870,245	1967	Bratley/McKay
57	2,082,464	2,090,656	1747	Euler
58	2,236,570	2,429,030	1966	Lee
59	2,652,728	2,941,672	1921	Mason
60	2,723,792	2,874,064	1929	Poulet
61	2,728,726	3,077,354	1966	Lee
62	2,739,704	2,928,136	1747	Euler
63	2,802,416	2,947,216	1747	Euler
64	2,803,580	3,716,164	1967	Bratley/McKay
65	3,276,856	3,721,544	1747	Euler
66	3,606,850	3,892,670	1967	Bratley/McKay
67	3,786,904	4,300,136	1747	Euler

	First number	Second number	Year of discovery	Discoverer
68	3,805,264	4,006,736	1929	Poulet
69	4,238,984	4,314,616	1967	Bratley/McKay
70	4,246,130	4,488,910	1747	Euler
71	4,259,750	4,445,050	1966	Lee
72	4,482,765	5,120,595	1957	Garcia
73	4,532,710	6,135,962	1957	Garcia
74	4,604,776	5,162,744	1966	Lee
75	5,123,090	5,504,110	1966	Lee
76	5,147,032	5,843,048	1747	Euler
77	5,232,010	5,799,542	1967	Bratley/McKay
78	5,357,625	5,684,679	1966	Lee
79	5,385,310	5,812,130	1967	Bratley/McKay
80	5,459,176	5,495,264	1967	Lee
81	5,726,072	6,369,928	1921	Mason
82	5,730,615	6,088,905	1966	Lee
83	5,864,660	7,489,324	1967	Bratley/McKay
84	6,329,416	6,371,384	1966	Lee
85	6,377,175	6,680,025	1966	Lee
86	6,955,216	7,418,864	1946	Escott
87	6,993,610	7,158,710	1957	Garcia
88	7,275,532	7,471,508	1967	Bratley/McKay
89	7,288,930	8,221,598	1966	Lee
90	7,489,112	7,674,088	1966	Lee
91	7,577,350	8,493,050	1966	Lee
92	7,677,248	7,684,672	1884	Seelhoff
93	7,800,544	7,916,696	1929	Gerardin
94	7,850,512	8,052,488	1966	Lee
95	8,262,136	8,369,864	1966	Lee
96	8,619,765	9,627,915	1957	Garcia
97	8,666,860	10,638,356	1966	Lee
98	8,754,130	10,893,230	1946	Escott
99	8,826,070	10,043,690	1967	Bratley/McKay
100	9,071,685	9,498,555	1946	Escott
101	9,199,496	9,592,504	1929	Gerardin/Poulet
102	9,206,925	10,791,795	1967	Bratley/McKay
103	9,339,704	9,892,936	1966	Lee
104	9,363,584	9,437,056	ca. 1600/1638	Yazdi/Descartes
105	9,478,910	11,049,730	1967	Bratley/McKay
106	9,491,625	10,950,615	1967	Bratley/McKay
107	9,660,950	10,025,290	1966	Lee
108	9,773,505	11,791,935	1967	Bratley/McKay

Figure 1-14

OTHER TYPES OF "FRIENDLY" NUMBERS

We can always look for nice relationships between numbers. Some of them are truly mind-boggling! Take, for example, the pair of numbers 6,205 and 3,869.

Do the following to verify these fantastic results.

$6,205 = 38^2 + 69^2$ and $3,869 = 62^2 + 05^2$
$5,965 = 77^2 + 06^2$ and $7,706 = 59^2 + 65^2$

And analogously, notice the symmetry:

$244 = 1^3 + 3^3 + 6^3$ and $136 = 2^3 + 4^3 + 4^3$

The search for other numbers related in some meaningful way can be challenging and rewarding. You might want to see if you can create other friendship pairs.

ANOTHER UNUSUAL NUMBER RELATIONSHIP

Can you imagine there are certain number pairs that yield the same product even when both numbers are reversed. For example, if $12 \cdot 42 = 504$ and if we reverse each of the two numbers, we get $21 \cdot 24 = 504$. The same thing is true for the number pair 36 and 84, since $36 \cdot 84 = 3,024 = 63 \cdot 48$.

At this point you may wonder if this will happen with any pair of numbers. The answer is that it will only work with fourteen pairs of numbers:

$$12 \cdot 42 = 21 \cdot 24 = 504$$
$$12 \cdot 63 = 21 \cdot 36 = 756$$
$$12 \cdot 84 = 21 \cdot 48 = 1,008$$
$$13 \cdot 62 = 31 \cdot 26 = 806$$
$$13 \cdot 93 = 31 \cdot 39 = 1,209$$
$$14 \cdot 82 = 41 \cdot 28 = 1,148$$
$$23 \cdot 64 = 32 \cdot 46 = 1,472$$
$$23 \cdot 96 = 32 \cdot 69 = 2,208$$
$$24 \cdot 63 = 42 \cdot 36 = 1,512$$
$$24 \cdot 84 = 42 \cdot 48 = 2,016$$
$$26 \cdot 93 = 62 \cdot 39 = 2,418$$
$$34 \cdot 86 = 43 \cdot 68 = 2,924$$
$$36 \cdot 84 = 63 \cdot 48 = 3,024$$
$$46 \cdot 96 = 64 \cdot 69 = 4,416$$

A careful inspection of these fourteen pairs of numbers will reveal that in each case the product of the tens digits of each pair of numbers is equal to the product of the units digits. You can also see this algebraically as follows for the numbers z_1, z_2, z_3, and z_4:

$$z_1 \cdot z_2 = (10a + b) \cdot (10c + d) = 100ac + 10ad + 10bc + bd$$
and
$$z_3 \cdot z_4 = (10b + a) \cdot (10d + c) = 100bd + 10bc + 10ad + ac$$

where a, b, c, d represent any of the ten digits: 0, 1, 2, . . . , 9, where $a \neq 0$ and $c \neq 0$.

We must have $z_1 \cdot z_2 = z_3 \cdot z_4$

$$
\begin{aligned}
100ac + 10ad + 10bc + bd &= 100bd + 10bc + 10ad + ac \\
100ac + bd &= 100bd + ac \\
99ac &= 99bd \\
ac &= bd
\end{aligned}
$$

which we observed earlier.

PERFECT NUMBERS

Most mathematics teachers probably told you often enough that everything in mathematics is perfect. While we would then assume that everything in mathematics is perfect, might there still be anything more perfect than something else? This brings us to numbers that hold such a title: *perfect numbers*. This is an official designation by the mathematics community. In the field of number theory, we have an entity called a "perfect number." This is defined as a number equal to the sum of its proper divisors (i.e., all the divisors except the number itself). The smallest perfect number is 6, since $6 = 1 + 2 + 3$, which is the sum of all its divisors excluding the number 6 itself.[17]

The next larger perfect number is 28, since $28 = 1 + 2 + 4 + 7 + 14$. And the next perfect number is $496 = 1 + 2 + 4 + 8 + 16 + 31 + 62 + 124 + 248$, which is the sum of all of the divisors of 496.

The first four perfect numbers were known to the ancient Greeks. They are: 6, 28, 496, and 8,128.

It was Euclid (ca. 300 BCE) who came up with a theorem to generalize a procedure to find a perfect number. He said that for an integer, k, if $2^k - 1$ is a prime number, then $2^{k-1}(2^k - 1)$ is a perfect number.

---

17. It is also the only number that is the sum and product of the same three numbers: $6 = 1 \cdot 2 \cdot 3 = 3!$ Also $6 = \sqrt{1^3 + 2^3 + 3^3}$. It is also fun to notice that $\frac{1}{1} = \frac{1}{2} + \frac{1}{3} + \frac{1}{6}$. By the way, while on the number 6, it is nice to realize that both 6 and its square, 36, are triangular numbers (see page 42).

This is to say that whenever we find a value of k that gives us a prime for $2^k - 1$, then we can construct a perfect number. We do not have to use all values of k, since if k is a composite number, then $2^k - 1$ is also a composite number.[18]

Using Euclid's method for generating perfect numbers we get the following table (figure 1-15):

Values of k	Values of $2^{k-1}(2^k - 1)$ when $2^k - 1$ is a prime number
2	6
3	28
5	496
7	8,128
13	33,550,336
17	8,589,869,056
19	137,438,691,328

Figure 1-15

The next few perfect numbers are:

2,305,843,008,139,952,128

2,658,455,991,569,831.7⁴´

191,561,942,60⁰

169,216

We ⁴

m to end in

triangular

$+ 3 + 4 +$

series

1⁵

4

8,

18. If $k =$ ₁ $= (2^p - 1)(2^{p(q-1)} + 2^{p(q-2)} + \ldots + 1)$. Therefore, $2^k - 1$ can only be prime when k is p₁ ... does not guarantee that when k is prime $2^k - 1$ will also be prime, as can be seen from the foll...ing values of k:

k	2	3	5	7	11	13
$2^k - 1$	3	7	31	127	2,047	8,191

where $2,047 = 23 \cdot 89$ is not a prime and so doesn't qualify.

This connection between the perfect numbers greater than 6 and the sum of the cubes of consecutive odd numbers is far more than could ever be expected!

You might try to find the partial sums for the next perfect numbers.

We do not know if there are any odd perfect numbers, since none has yet been found. Using today's computers, we have a much greater facility at discovering more perfect numbers. You might try to find larger perfect numbers using Euclid's method.

TRIANGULAR NUMBERS

We can categorize numbers by the shape that can be formed by the number of squares they represent. For example, if we have four squares, we can easily form a square arrangement. Similarly, this can be done with nine squares or with sixteen squares. These are well known as square numbers: 1, 4, 9, 16, 25, 36, 49, . . . (see figure 1-16), which evolve (as n^2) from the sum of the first n odd natural numbers.

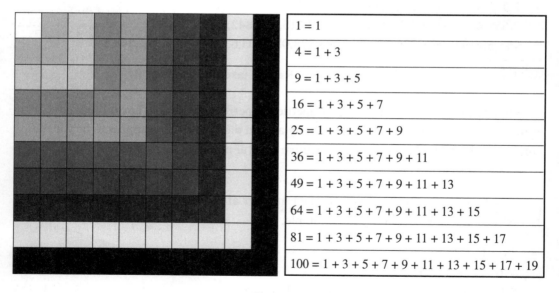

1 = 1
4 = 1 + 3
9 = 1 + 3 + 5
16 = 1 + 3 + 5 + 7
25 = 1 + 3 + 5 + 7 + 9
36 = 1 + 3 + 5 + 7 + 9 + 11
49 = 1 + 3 + 5 + 7 + 9 + 11 + 13
64 = 1 + 3 + 5 + 7 + 9 + 11 + 13 + 15
81 = 1 + 3 + 5 + 7 + 9 + 11 + 13 + 15 + 17
100 = 1 + 3 + 5 + 7 + 9 + 11 + 13 + 15 + 17 + 19

Figure 1-16

What is less well known, but perhaps more enchanting, are *triangular numbers*. These are numbers representing dots that can form an equilateral triangle, as shown in figure 1-17. They are: 1, 3, 6, 10, 15, 21, 28, ...

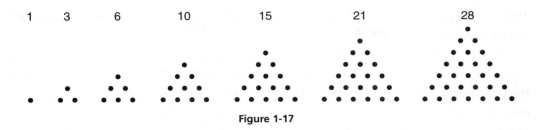

Figure 1-17

The nth triangular number, which we will call t_n, can be obtained quite simply by taking the sum of the first n natural numbers (see figure 1-18 and the appendix).

n	Sum of the first n natural numbers	Triangular numbers t_n
1	1	1
2	1 + 2	3
3	1 + 2 + 3	6
4	1 + 2 + 3 + 4	10
5	1 + 2 + 3 + 4 + 5	15
6	1 + 2 + 3 + 4 + 5+ 6	21
7	1 + 2 + 3 + 4 + 5+ 6 + 7	28
8	1 + 2 + 3 + 4 + 5+ 6 + 7 + 8	36
9	1 + 2 + 3 + 4 + 5+ 6 + 7 + 8 + 9	45
10	1 + 2 + 3 + 4 + 5+ 6 + 7 + 8 + 9 + 10	55
11	1 + 2 + 3 + 4 + 5+ 6 + 7 + 8 + 9 + 10 + 11	66
12	1 + 2 + 3 + 4 + 5+ 6 + 7 + 8 + 9 + 10+ 11 + 12	78
13	1 + 2 + 3 + 4 + 5+ 6 + 7 + 8 + 9 + 10 + 11 + 12 + 13	91
14	1 + 2 + 3 + 4 + 5+ 6 + 7 + 8 + 9 + 10 + 11 + 12 + 13 + 14	105
15	1 + 2 + 3 + 4 + 5+ 6 + 7 + 8 + 9 + 10 + 11 + 12 + 13 + 14 + 15	120
16	1 + 2 + 3 + 4 + 5+ 6 + 7 + 8 + 9 + 10 + 11 + 12 + 13 + 14 + 15 + 16	136
17	1 + 2 + 3 + 4 + 5+ 6 + 7 + 8 + 9 + 10 + 11 + 12 + 13 + 14 + 15 + 16 + 17	153
18	1 + 2 + 3 + 4 + 5+ 6 + 7 + 8 + 9 + 10 + 11 + 12 + 13 + 14 + 15+ 16 + 17 + 18	171

Figure 1-18

Let's digress for a moment to develop a shortcut for adding consecutive natural numbers. Carl Friedrich Gauss (1777–1855) was perhaps one of the most gifted mathematicians of all time. Part of his success in mathematics can be attributed to his uncanny ability to "number crunch," that is, to do arithmetic calculations with incredible speed. Many of his theorems evolved from his ability to do what most others could not. It gave him a much greater insight into quantitative relationships and enabled him to make the many breakthroughs for which he is still so famous today. The often told story[19] about his days in

---

19. According to E.T. Bell in his famous book *Men of Mathematics* (Simon and Schuster, 1937), Gauss, in his adult years, told of this story but explained that the situation was far more complicated than the simple one we currently tell. He told of his teacher, Mr. Büttner, giving the class a five-digit number, such as 81,297, and they were asked to add a three-digit number such as 198 to it 100 times successively, and then find the sum of that series. One can only speculate about which version is the true one!

elementary school can be used as an excellent lead-in to the topic of adding an arithmetic sequence of numbers.

As the story goes, young Gauss's teacher, Mr. Büttner, wanted to keep the class occupied; so he asked the class to add the numbers from 1 to 100. He had barely finished giving the assignment when young Gauss put his slate down with simply one number on it, the correct answer! Of course, Mr. Büttner assumed Gauss had the wrong answer, or cheated. In any case, he ignored this response and waited for the appropriate time to ask the students for their answers. No one, other than Gauss, had the right answer. What did Gauss do to get the answer mentally? Gauss explained his method: rather than add the numbers in the order in which they appear, $1 + 2 + 3 + 4 + \ldots + 97 + 98 + 99 + 100$, he felt it made more sense to add the first and the last, then add the second and the next-to-last, then the third and the third-from-the-last, and so on. This led to a much simpler addition:

$1 + 100 = 101$; $2 + 99 = 101$; $3 + 98 = 101$; $4 + 97 = 101$

More simply put, we now have fifty pairs of numbers with the sum of each pair being 101; or $50 \cdot 101 = 5,050$. In effect, he established the formula that the sum of the consecutive natural numbers $S = \frac{n}{2} \cdot (a + b)$, where n is the number of numbers being added, a is the first number, and b is the last number.

To use Gauss's method to find the sum of the first n natural numbers, we find the sum of the first and last numbers, then multiply this sum by half the number of numbers being added. For example, if we want to find the two hundredth triangular number, which we will denote by t_{200}, we would have to find the sum of the natural numbers from 1 to 200: $1 + 2 + 3 + 4 + 5 + \ldots + 196 + 197 + 198 + 199 + 200$. Taking their sum in pairs we get one hundred pairs of 201 (i.e., $1 + 200 = 201$, $2 + 199 = 201$, $\ldots$) for a sum of $20,100 = t_{200}$.

In general, we could use the following formula to find the nth triangular number:

$$t_n = \frac{n}{2} \cdot (1 + n) = \frac{n \cdot (n + 1)}{2}$$

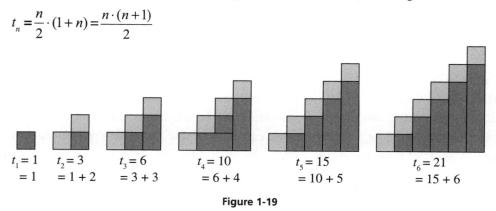

$t_1 = 1$ $\quad$ $t_2 = 3$ $\quad$ $t_3 = 6$ $\quad$ $t_4 = 10$ $\quad$ $t_5 = 15$ $\quad$ $t_6 = 21$
$= 1$ $\quad\quad$ $= 1 + 2$ $\quad$ $= 3 + 3$ $\quad$ $= 6 + 4$ $\quad$ $= 10 + 5$ $\quad$ $= 15 + 6$

Figure 1-19

There is a plethora of number curiosities involving triangular numbers, which is what makes them so spectacular. We offer some here to show you the vast reaches of these numbers.

- The sum of any two consecutive triangular numbers is always a square number. For example, $6 + 10 = 16 = 4^2$ and $21 + 28 = 49 = 7^2$. Symbolically, we could write this as: $t_n + t_{n+1} =$ a square number. You might want to justify this with dot arrangements by showing that when the triangular dot arrangements of two consecutive triangular numbers are placed beside one another, a square arrangement can be made (see figure 1-20).

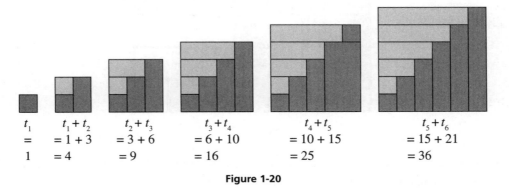

$$t_1 \qquad t_1 + t_2 \qquad t_2 + t_3 \qquad t_3 + t_4 \qquad t_4 + t_5 \qquad t_5 + t_6$$
$$= \qquad = 1 + 3 \qquad = 3 + 6 \qquad = 6 + 10 \qquad = 10 + 15 \qquad = 15 + 21$$
$$1 \qquad = 4 \qquad = 9 \qquad = 16 \qquad = 25 \qquad = 36$$

Figure 1-20

- A triangular number will never have a 2, 4, 7, or 9 as a terminal digit.

- If you multiply a triangular number by 9 and then add 1, you will always end up with another triangular number. For example:

$$9 \cdot t_3 + 1 = 9 \cdot 6 + 1 = 55 = t_{10}$$
$$9 \cdot t_4 + 1 = 9 \cdot 10 + 1 = 91 = t_{13}$$
$$9 \cdot t_7 + 1 = 9 \cdot 28 + 1 = 253 = t_{22}$$

Symbolically, we could write this as $9t_n + 1 = t_{3n+1}$, a triangular number.

- If you multiply a triangular number by 8 and then add 1, you will always end up with a square number. For example:

$$8 \cdot t_3 + 1 = 8 \cdot 6 + 1 = 49 = 7^2 = s_7$$
$$8 \cdot t_5 + 1 = 8 \cdot 15 + 1 = 121 = 11^2 = s_{11}$$
$$8 \cdot t_8 + 1 = 8 \cdot 36 + 1 = 289 = 17^2 = s_{17}$$

Symbolically, we could write this as $8t_n + 1 = s_{2n+1}$, a square number.

Geometrically you can see the justification for this conclusion since we can arrange geometrically eight copies of a triangular number (in the middle) plus an extra unit to form a square as shown in figure 1-21.

Figure 1-21

- The sum of the cubes of the first n natural numbers is equal to the square of the nth triangular number. For example:

$$1^3 + 2^3 + 3^3 + 4^3 + 5^3 = 1 + 8 + 27 + 64 + 125 = 225 = 15^2 = t_5^2$$

which is the fifth triangular number squared.

Symbolically, we could write this as $t_n^2 = 1^3 + 2^3 + 3^3 + 4^3 + \cdots + n^3$.

- For any natural number n, if we take the sum of the powers of 9 to the nth number, we will arrive at a triangular number. That is, $9^0 + 9^1 + 9^2 + 9^3 + 9^4 + \cdots + 9^n$ will always be a triangular number. Let's look at the first few of these values to convince ourselves that this "works":

$$9^0 = 1 = t_1$$
$$9^0 + 9^1 = 10 = t_4$$
$$9^0 + 9^1 + 9^2 = 91 = t_{13}$$
$$9^0 + 9^1 + 9^2 + 9^3 = 820 = t_{40}$$
$$9^0 + 9^1 + 9^2 + 9^3 + 9^4 = 7,381 = t_{121}$$

- Every fourth power of a number greater than 1 can be expressed as the sum of two triangular numbers:

$$2^4 = 16 = s_4 = 6 + 10 = t_3 + t_4$$
$$3^4 = 81 = s_9 = 36 + 45 = t_8 + t_9$$
$$4^4 = 256 = s_{16} = 120 + 136 = t_{15} + t_{16}$$

You may see the pattern that is evolving: the number taken to the fourth power, when squared, is the subscript (or number of) the larger of the two triangular numbers of the sum. Symbolically, we could write this as: $n^4 = s_n^2 = t_{n^2-1} + t_{n^2}$.

- A triangular number greater than 1 cannot be a number to the third, fourth, or fifth power.

- All perfect numbers are triangular numbers: 6, 28, 495, 8,128, 33,550,336 ...
  $6 = t_3$, $28 = t_7$, $495 = t_{31}$, $8,128 = t_{127}$, $33,550,336 = t_{8,191}$, ...

- The number 3 is the only triangular number that is a prime number.

- We can write all natural numbers as the sum of at most three triangular numbers. Here are a few examples: $7 = 6 + 1$; $23 = 10 + 10 + 3$; and $30 = 21 + 6 + 3$. You might want to try to express some other numbers as the sum of at most three triangular numbers.[20]

- There are twenty-eight triangular numbers less than 10^{10} that are palindromic: Here they are:

  $1 = t_1$
  $3 = t_2$
  $6 = t_3$
  $55 = t_{10}$
  $66 = t_{11}$ [the index 11 is also a palindrome][21]
  $171 = t_{18}$
  $595 = t_{34}$
  $666 = t_{36}$ [see page 12]
  $3,003 = t_{77}$ [index is also a palindrome]
  $5,995 = t_{109}$
  $8,778 = t_{132}$
  $15,051 = t_{173}$
  $66,066 = t_{363}$ [index is also a palindrome]
  $617,716 = t_{1,111}$ [index is also a palindrome]
  $828,828 = t_{1,287}$

---

20. The famous mathematician Carl Friedrich Gauss was extremely proud to discover this relationship. On July 10, 1796, he wrote in his diary, "EYPHKA [=Eureka]! num = Δ + Δ + Δ." In other words, each number (num) is the sum of at most three triangular numbers (Δ).

21. The subscript that indicates the number (or position) of the triangular number we call its *index*.

$$1,269,621 = t_{1,593}$$
$$1,680,861 = t_{1,833}$$
$$3,544,453 = t_{2,662} \text{ [index is also a palindrome]}$$
$$5,073,705 = t_{3,185}$$
$$5,676,765 = t_{3,369}$$
$$6,295,926 = t_{3,548}$$
$$35,133,153 = t_{8,382}$$
$$61,477,416 = t_{11,088}$$
$$178,727,871 = t_{18,906}$$
$$1,264,114,621 = t_{50,281}$$
$$1,634,004,361 = t_{57,166}$$
$$5,289,009,825 = t_{102,849}$$
$$6,172,882,716 = t_{111,111} \text{ [index is also a palindrome]}$$

- Although all palindromic triangular numbers can be reversed to remain as triangular numbers, there are some triangular numbers that, when reversed, are also triangular numbers. For example, the following triangular numbers, when reversed, are also triangular numbers (the palindromic triangular numbers are underlined):

  1; 3; 6; 10; 55; 66; 120; 153; 171; 190; 300; 351; 595; 630; 666; 820; 3,003; 5,995; 8,778; 15,051; 17,578; 66,066; 87,571; 156,520; 180,300; 185,745; 547,581; 557,040; 617,716; 678,030; 828,828; 1,269,621; 1,461,195; 1,680,861; 1,851,850; 3,544,453; 5,073,705; 5,676,765; 5,911,641; 6,056,940; 6,295,926; 12,145,056; 2,517,506; 16,678,200; 35,133,153; 56,440,000; 60,571,521; 61,477,416; 65,054,121; 157,433,640; 178,727,871; 188,267,310; 304,119,453; 354,911,403; 1,261,250,200; 1,264,114,621; 1,382,301,910; 1,634,004,361; 1,775,275,491; and 1,945,725,771

- Some triangular numbers are also square numbers, such as:

$$t_1 = 1 = 1^2 = s_1$$
$$t_8 = 36 = 6^2 = s_6$$
$$t_{49} = 1,225 = 35^2 = s_{35}$$
$$t_{288} = 41,616 = 204^2 = s_{204}$$
$$t_{1,681} = 1,413,721 = 1,189^2 = s_{1,189}$$
$$t_{9,800} = 48,024,900 = 6,930^2 = s_{6,930}$$
$$t_{57,121} = 1,631,432,881 = 40,391^2 = s_{40,391}$$
$$t_{332,928} = 55,420,693,056 = 235,416^2 = s_{235,416}$$

- Some pairs of triangular numbers have a sum and a difference that are also triangular numbers (see figure 1-22):

The pair of triangular numbers		Their sum		Their difference	
$t_5 = 15,$	$t_6 = 21$	$15 + 21 = 36$	$= t_8$	$21 - 15 = 6$	$= t_3$
$t_{14} = 105,$	$t_{18} = 171$	$105 + 171 = 276$	$= t_{23}$	$171 - 105 = 66$	$= t_{11}$
$t_{27} = 378,$	$t_{37} = 703$	$378 + 703 = 1{,}081$	$= t_{46}$	$703 - 378 = 325$	$= t_{25}$
$t_{39} = 780,$	$t_{44} = 990:$	$t_{44} + t_{39} = 1{,}770$	$= t_{59}$	$t_{44} - t_{39} = 210$	$= t_{20}$
$t_{54} = 1{,}485,$	$t_{91} = 4{,}186:$	$t_{91} + t_{54} = 5{,}671$	$= t_{106}$	$t_{91} - t_{54} = 2{,}701$	$= t_{73}$
$t_{65} = 2{,}145,$	$t_{86} = 3{,}741:$	$t_{86} + t_{65} = 5{,}886$	$= t_{108}$	$t_{86} - t_{65} = 1{,}596$	$= t_{56}$
$t_{104} = 5{,}460,$	$t_{116} = 6{,}786:$	$t_{116} + t_{104} = 12{,}246$	$= t_{156}$	$t_{116} - t_{104} = 1{,}326$	$= t_{51}$
$t_{125} = 7{,}875,$	$t_{132} = 8{,}778:$	$t_{132} + t_{125} = 16{,}653$	$= t_{182}$	$t_{132} - t_{125} = 903$	$= t_{42}$

Figure 1-22

- Some triangular numbers are the product of two consecutive numbers. Figure 1-23 shows some of these:

$$2 \cdot 3 = 6 = t_3$$
$$14 \cdot 15 = 210 = t_{20}$$
$$84 \cdot 85 = 7{,}140 = t_{119}$$
$$492 \cdot 493 = 242{,}556 = t_{696}$$
$$2{,}870 \cdot 2{,}871 = 8{,}239{,}770 = t_{4{,}059}$$
$$16{,}730 \cdot 16{,}731 = 279{,}909{,}630 = t_{23{,}660}$$
$$97{,}512 \cdot 97{,}513 = 9{,}508{,}687{,}656 = t_{137{,}903}$$

Figure 1-23

- Six triangular numbers are the product of three consecutive numbers. They are (figure 1-24):

$$1 \cdot 2 \cdot 3 = 6 = t_3$$
$$4 \cdot 5 \cdot 6 = 120 = t_{15}$$
$$5 \cdot 6 \cdot 7 = 210 = t_{20}$$
$$9 \cdot 10 \cdot 11 = 990 = t_{44}$$
$$56 \cdot 57 \cdot 58 = 185{,}136 = t_{608}$$
$$636 \cdot 637 \cdot 638 = 258{,}474{,}216 = t_{22{,}736}$$

Figure 1-24

- As a matter of fact, there is even a triangular number that is the product of three, four, and five consecutive numbers: $4 \cdot 5 \cdot 6 = 2 \cdot 3 \cdot 4 \cdot 5 = 1 \cdot 2 \cdot 3 \cdot 4 \cdot 5 = 120 = t_{15} = 5!$

- Some triangular numbers are the product of two prime numbers. Some of these are (figure 1-25):

$$
\begin{array}{rcl}
2 \cdot 3 & = & 6 & = & t_3 \\
3 \cdot 5 & = & 15 & = & t_5 \\
3 \cdot 7 & = & 21 & = & t_6 \\
5 \cdot 11 & = & 55 & = & t_{10} \\
7 \cdot 13 & = & 91 & = & t_{13} \\
11 \cdot 23 & = & 253 & = & t_{22} \\
19 \cdot 37 & = & 703 & = & t_{37}
\end{array}
$$

Figure 1-25

- Some triangular numbers are divisible by the sum of their digits. For example: the triangular number $t_{17} = 153$ is divisible by the sum of its digits: $1 + 5 + 3 = 9$, since $\frac{153}{9} = 17$. Other triangular numbers that are divisible by the sum of their digits are:

1; 3; 6; 10; 21; 36; 45; 120; 153; 171; 190; 210; 300; 351; 378; 465; 630; 666; 780; 820; 990; 1,035; 1,128; 1,275; 1,431; 1,540; 1,596; 1,770; 2,016; 2,080; 2,556; 2,628; 2,850; 2,926; 3,160; 3,240; 3,321; 3,486; 3,570; 4,005; 4,465; 4,560; 4,950; 5,050; 5,460; 5,565; 5,778; 5,886; 7,140; 7,260; 8,001; 8,911; 9,180; 10,011; 10,296; 10,440; 11,175; 11,476; 11,628; 12,720; 13,041; 13,203; 14,196; 14,706; 15,225; 15,400; 15,576; 16,110; 16,290; 16,653; 17,020; 17,205; 17,766; 17,955; 18,145; 18,528; 20,100; 21,321; 21,528; 21,736; 21,945; 22,155; 23,220; 23,436; 24,090; 24,310; 24,976; 25,200; 28,680; 29,646; 30,628; 31,626; 32,640; 33,930; 35,245; 36,585; 37,128; 39,060; 40,470; 41,328; 41,616; 43,365; 43,956; 45,150; 46,360; 51,040; 51,360; 51,681; 52,326; 52,650; 53,956; 56,280; 56,616; 61,776; 63,903; 64,620; 65,341; 67,896; 69,006; 70,125; 70,500; 72,010; 73,536; 73,920; 76,636; 78,210; 79,401; 79,800; 80,200; 81,810; 88,410; 89,676; 90,100; 93,096; 93,528; 97,020; 100,128; 101,025; 103,740; 105,111

The following table (figure 1-26) shows the first few of the triangular numbers, t_n, as they relate to the sum of their digits, $S(t_n)$:

n	t_n	$S(t_n)$	$t_n / S(t_n)$	n	t_n	$S(t_n)$	$t_n / S(t_n)$
1	1	1	1	26	351	9	39
2	3	3	1	27	378	18	21
3	6	6	1	30	465	15	31
4	10	1	10	35	630	9	70
6	21	3	7	36	**666**	18	37
8	36	9	4	39	780	15	52
9	45	9	5	40	820	10	82
15	120	3	40	44	990	18	55
17	153	9	17	45	1,035	9	115
18	171	9	19	47	1,128	12	94
19	190	10	19	50	1,275	15	85
20	210	3	70	53	1,431	9	159
24	300	3	100	55	1,540	10	154

Figure 1-26

- The only triangular numbers that are repeating-digit numbers are: $t_{10} = 55$, $t_{11} = 66$, and $t_{36} = 666$. (Remember this last one from our earlier discussion?)

- The product of some pairs of triangular numbers results in a square number. For example (figure 1-27):

$$t_2 \cdot t_{24} = \quad 3 \cdot 300 = \quad\quad 900 = \quad 30^2 = s_{30}$$
$$t_2 \cdot t_{242} = \quad 3 \cdot 29{,}403 = \quad 88{,}209 = 297^2 = s_{297}$$
$$t_3 \cdot t_{48} = \quad 6 \cdot 1{,}176 = \quad\quad 7{,}056 = \quad 84^2 = s_{84}$$
$$t_6 \cdot t_{168} = 21 \cdot 14{,}196 = 298{,}116 = 546^2 = s_{546}$$

Figure 1-27

- Triangular numbers also can be shown to form some nice patterns. Consider the following (figure 1-28) nifty little pattern:

$$t_1 + t_2 + t_3 = \quad\quad 10 = \quad t_4$$
$$t_5 + t_6 + t_7 + t_8 = \quad\quad 100 = \quad t_9 + t_{10}$$
$$t_{11} + t_{12} + t_{13} + t_{14} + t_{15} = \quad\quad 460 = \quad t_{16} + t_{17} + t_{18}$$
$$t_{19} + t_{20} + t_{21} + t_{22} + t_{23} + t_{24} = \quad 1{,}460 = \quad t_{25} + t_{26} + t_{27} + t_{28}$$

Figure 1-28

There is even a Pythagorean triple[22] comprised of triangular numbers, so that
$$t_{132}^2 + t_{143}^2 = 8{,}778^2 + 10{,}296^2 = 183{,}060{,}900 = 13{,}530^2 = t_{164}^2 .$$

In the famous Pascal triangle[23] the triangular numbers appear as one of the diagonals. See the shaded cells in figure 1-29.

The Pascal Triangle

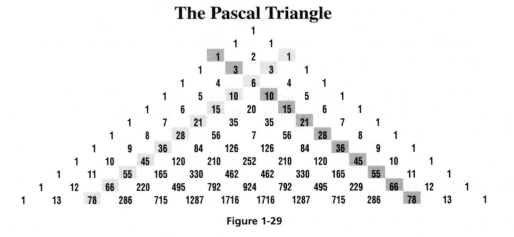

Figure 1-29

22. A *Pythagorean triple* is a set of three numbers where the sum of the squares of two of them is equal to the square of the third number.
23. The *Pascal triangle*, named after the famous French mathematician Blaise Pascal (1623–1662), is a triangular arrangement of numbers where the sum of two adjacent numbers is placed below and between them.

SQUARES THAT END IN THE ORIGINAL NUMBER

We know that when we square some numbers, the resulting number will end with the original number. For example, when we square the number 5, we get 25, whose terminal digit is a 5. Similarly, when we square 6, we get 36, whose terminal digit is a 6. With a two-digit starting number, consider squaring 25, we get 625, where the 25 reappears at the end of the number. Another such number is 76, for when we square it we get 5,776, again ending with the same number as the one we originally squared. The next such number is 100, which you can readily verify. If we use a computer to search for the three-digit numbers that have this property, we would find only two numbers whose square ends in the same number as that with which we started. They are:

$$376^2 = 141,376$$
$$625^2 = 390,625$$

Here is a table (figure 1-30) that shows the squares that end with their starting number (omitting those with terminal zeros) between 0 and 300 million:

n	n^2	n	n^2
1	1	90,625	8,212,890,625
5	25	109,376	11,963,109,376
6	36	890,625	793,212,890,625
25	625	2,890,625	8,355,712,890,625
76	5,776	7,109,376	50,543,227,109,376
376	141,376	12,890,625	166,168,212,890,625
625	390,625	87,109,376	7,588,043,387,109,376
9,376	87,909,376	212,890,625	45,322,418,212,890,625

Figure 1-30

You might want to inspect this list of squares and see if you can determine a pattern among those that end in a 5 and those that end in a 6. Here, again, we see the beauty in mathematics that is often hidden.

NUMEROLOGY

Before we close our discussion of numbers, we could provide some insights from a less mathematical viewpoint. Numerology inspects numbers with other points of view, such as those of psychology or philosophy. In 1995, a study[24] concluded that the easiest numbers

---

24. Marisca Milikowski and Jan J. Elshout, "What Makes Numbers Easy to Remember?" *British Journal of Psychology* (1995): 537–47.

to remember up to 100 were 8, 1, 100, 2, 17, 5, 9, 10, 99, and 11, while the most difficult numbers to remember were 82, 56, 61, 94, 85, 45, 83, 59, 41, and 79.

This psychological study need not be relevant to you but may be for others!

Then there is the constant superstition about the number 13. Some contend that it has many unpleasant associations in the Bible, and others are more concerned if it turns up as the date of a Friday, which it tends to do more frequently than any other day of the week;[25] perhaps causing some to find that to be an unlucky day. Then there are well-known triskaidekaphobics,[26] such as Napoleon Bonaparte, Herbert Hoover, Mark Twain, Richard Wagner, and Franklin Delano Roosevelt. Take for one, Richard Wagner (1813–1883), the famous German composer who revolutionized music. Here we can see how, for him, the number 13 has played a curious role:

Wagner was born in 18**13**, which has a digit sum of **13**.
Wagner died on February **13**, 1883, which was the **13**th year of the unification of Germany.
Wagner wrote **13** operas during his lifetime.
Richard Wagner's name consists of **13** letters.
Wagner's opera *Tannhäuser* was completed on April **13**, 1845.
On May **13**, 1861, Wagner premiered *Tannhäuser* in Paris during his **13**-year exile from Germany.
The grand opening of Wagner's festival opera house in Bayreuth, Germany, was opened on August **13**, 1876.
Wagner completed his last opera, *Parsifal*, on January **13**, 1882.
Wagner's last day in Bayreuth, the city in which he built his famous festival opera house, was September **13**, 1882.

This kind of "numerology" is just for entertainment, and does not involve mathematics. Yet we thought we would close this chapter on a lighter note. Having journeyed through a vast array of numbers, we hope you are feeling a certain degree of awe from the many amazing properties you have witnessed in our decimal system. There are many more such hidden beauties, and we leave them for you to discover.

Who said numbers can't form beautiful relationships! Having experienced some of these unique situations might give you the feeling that there is more to "numbers" than meets the eye. While, rightfully, you may want to verify these relationships, we encourage you to find others that can also be considered "beautiful."

---

25. See A. S. Posamentier, *Math Charmers: Tantalizing Tidbits for the Mind* (Amherst, NY: Prometheus Books, 2003), p. 230.
26. A phobia for the number 13.

Chapter 2

ARITHMETIC ENTERTAINMENTS
AND NOVELTIES

I n recent years, the art of doing arithmetic has been somewhat pushed aside by the advent of the calculator and computer. Yet, by experiencing arithmetic relationships and the many hidden treasures within arithmetic possesses, one gets a much firmer grasp of numbers and a greater appreciation of mathematics. Yes, we will actually make arithmetic fun! Let us first consider those whose arithmetic skills seemed supernatural.

CALCULATING GENIUSES

One of the earliest calculating geniuses was Zerah Colburn (1804–1839). He was the son of a farmer from Vermont. Even before he learned to read, he was able to multiply two numbers up to 100. His father toured him through the United States so that he could finance his education in London and Paris. At age eight, his genius was acknowledged in London in the *Annual Register* as the most unusual phenomenon in the history of humankind. In his autobiography, he mentions one of the challenges presented to him in New Hampshire in 1811. There, he was asked to calculate the number of days and hours from the time of Jesus to the beginning of 1811. It took him twenty seconds to respond 661,015 days or 15,864,360 hours. When asked how many seconds there are in eleven years, he answered, within four seconds: 346,896,000. When Colburn did multiplication, he tended to see numbers in terms of their factors, so for him to multiply 21,734 by 543, he would factor 543 into 181 and 3, then he would multiply 21,734 by 181 to get 3,933,854, and then multiply that by 3 to get 11,801,562.

In a similar situation, the British novelty George Parker Bidder (1806–1878) also traveled with his father on tour showing off his talents before the age of nine. He, for example, was able to find at age ten $\sqrt{119,550,669,121} = 345,761$. In 1818, these two young fellows, Colburn and Bidder, competed in a contest and Bidder won. Bidder went on to graduate from the University of Edinburgh, where he studied engineering, while Colburn abandoned his lightning-fast calculating skills and became a Methodist minister.

In Germany, Johann Martin Zacharias Dase (1824–1861) gained fame when, in 1861, he was able to multiply two 20-digit numbers correctly in 6 minutes, multiply two 48-digit numbers correctly in 40 minutes, and multiply two 100-digit numbers in 8 ¾ hours. His ability to multiply two 8-digit numbers in 54 seconds was a record in his day. In 1839, Dase toured his talents in Germany, Austria, and England, yet he did not have a very sophisticated mathematical talent. Since he was such a lightning-fast calculator, the famous German mathematician Carl Friedrich Gauss (1777–1855) often used him to perform calculations, in essence using him as an automatic calculator. This was clearly a compliment, since Gauss was also known for his amazing facilities with numbers. One of the tasks that Gauss had him do was to factor into primes the natural numbers, which he ended up doing from 7 up to 9 million.

The Indian mathematician Srinivasa Ramanujan (1887–1920) is known in higher mathematics for having enormous mathematical powers that allowed him to come up with startling truths previously unknown. He worked closely with the famous English mathematician Godfrey Harold Hardy (1877–1947) at Trinity College at Cambridge University.

The number 1729, known as the *Hardy-Ramanujan number*, got its name and fame from an anecdote relayed by Hardy regarding a hospital visit to Ramanujan. In Hardy's words:

> I remember once going to see him when he was ill at Putney. I had ridden in taxi cab number 1729 and remarked that the number seemed to me rather a dull one, and that I hoped it was not an unfavorable omen. "No," he replied, "it is a very interesting number; it is the smallest number expressible as the sum of two cubes in two different ways."[1]

The quotation is sometimes expressed using the term "positive cubes," since the acceptance of negative perfect cubes (the cube of a negative integer) gives the smallest solution as 91 (which happens to be a factor of 1729), that is, $91 = 6^3 + (-5)^3 = 4^3 + 3^3$.

We then have that $1,729 = 1^3 + 12^3 = 1 + 1,728$ and $1,729 = 10^3 + 9^3 = 1,000 + 729$. This is the smallest number that can be expressed as the sum of two positive cubes in *two*

---

1. http://www-gap.dcs.st-and.ac.uk/~history/Quotations/Hardy.html.

different ways. You will notice that the smallest positive number that can be expressed as the sum of two positive cubes (in one way) is $2 = 1^3 + 1^3$.

New Zealand also contributed to this kind of genius with Alexander Craig Aitken (1895–1967), who was later a professor of mathematics at the University of Edinburgh. Lore has it that when asked to change the fraction $\frac{4}{47}$ to decimal form, he claimed the answer .0851063829787234042553191... and said that in four seconds that was the best he could do. He claimed that as soon as he purchased a mechanical calculator, his mental calculating speed began to diminish. What does this say about our use of electronic calculators today?

India contributed to speed calculation with Shakuntala Devi (1939–), who, in 1977, was able to find the twenty-third root of a 201-digit number. She also found her way into the *Guinness Book of World Records*, by multiplying 7,686,369,774,870 times 2,465,099,745,779 to get the correct answer, 18,947,668,177,995,426,462,773,730, in the unbelievable time of 28 seconds.

The first world championship of mental calculation took place in 2004 in the German city of Annaberg-Buchholz. The seventeen participants were challenged to add ten 10-digit numbers, multiply two 8-digit numbers, and take the square root of a 6-digit number, all within ten minutes. Britain's Robert Fountain was the winner. He attempted to defend his title two years later against twenty-six participants, but only won in the competition of root-extraction of numbers. In November 2007, twenty-seven-year-old Alexis Lemaire from France won the world championship for mental calculation in New York City, by taking the thirteenth root of a 200-digit number in 72.4 seconds. A year later, the title went to Spaniard Alberto Coto, who took first place against twenty-seven participants in Leipzig, Germany.

But dearest to us is our friend Arthur Benjamin (1961–), a professor of mathematics at Harvey Mudd College in California, who, although not a competitor for the world championships just cited, has used his mental calculation talent in a far more productive way: education. This master teacher brings to his audiences the mental algorithms[2] that he uses for his lightning-fast calculation, whether it's extracting a square root of a 5-digit number in the time it takes to write the number down or determining the day of the week on which a randomly selected person was born. He motivates audiences of students, teachers, and the general population, not only to popularize mathematics, but also to motivate their further study of it. This is a worthwhile endeavor.

DIGIT SUMS

When we speak of the sum of the digits of a number, we simply add the digits. For example, the sum of the digits of 251 is simply $2 + 5 + 1 = 8$.

---

2. We have defined an algorithm as a step-by-step problem-solving procedure, especially an established, recursive computational procedure for solving a problem in a finite number of steps.

Let's take a big leap and consider what the sum of the digits is of all the numbers from 1 to 1,000,000. One way to answer this question is to begin by summing the digits of consecutive numbers starting with 1 (see figure 2-1).

The number	The digit sum
1	1
2	2
3	3
...	...
35	$3 + 5 = 8$
36	$3 + 6 = 9$
37	$3 + 7 = 10$
38	$3 + 8 = 11$
etc.	

Figure 2-1

This does not seem to be a very efficient or elegant way of proceeding. To determine this sum, rather than simply trying to add the digits of all the numbers, we will use a more efficient arrangement of the numbers and see if we can obtain the sum without actually doing all this tedious addition.

Let's consider a list of the numbers from 0 to 999,999 in two directions: ascending order and descending order (see figure 2-2).

Ascending order	Descending order	Sum of the digits of each pair of numbers
0	999,999	54
1	999,998	54
2	999,997	54
3	999,996	54
...	...	...
127	999,872	54
...	...	...
257,894	742,105	54
...	...	...
999,997	2	54
999,998	1	54
999,999	0	54

Figure 2-2

We will leave the number 1,000,000 till a bit later.

Since every pair of numbers has a digit sum of 54, and the two columns have the same numbers, the sum of the digits in one column is $27 \cdot 1,000,000$. We must now add the digit sum of the last number, 1,000,000, which is 1.

Therefore, the digit sum of all numbers from 1 to 1,000,000 is $1,000,000 \cdot 27 + 1$ = 27,000,001. From this example, you can see that arithmetic is more than just doing the basic operations—addition, subtraction, multiplication, and division. It requires a bit of thinking as well.

HOW FAR CAN YOU COUNT?

When one is asked how far can one count, the answer is usually: "as far as I can identify the numbers." Yet simple multiplication can answer the question rather definitively. On average it takes about one second to speak out a two-digit number and about five seconds to say a six-digit number. So let's estimate that; say it takes on average four seconds to say a number. Suppose we begin our counting exercise at age ten and continue to age seventy—without any breaks!

Without calculating the leap years, we would reach the number 473,040,000 in the sixty years of continuous counting. This was reached by simple arithmetic. We will be counting for 60 years, each year has 365 days, each day has 24 hours, each hour has 60 minutes, and each minute has 60 seconds. This then results in $\frac{60 \cdot 365 \cdot 24 \cdot 60 \cdot 60}{4} = 473,040,000$, or just a bit over 473 million, and that's without any pauses!

DETERMINING THE LARGER SUM—WITHOUT ADDING

Consider the sum of the following numbers where each succeeding number has an additional digit in the sequence of natural numbers 1 to 9. In the first case, we set up the numbers in ascending order; and in the second case, we set up the numbers in descending order.

$S_1 = 1 + 12 + 123 + 1,234 + 12,345 + 123,456 + 1,234,567 + 12,345,678 + 123,456,789$
and
$S_2 = 987,654,321 + 87,654,321 + 7,654,321 + 654,321 + 54,321 + 4,321 + 321 + 21 + 1$

Which sum is larger, S_1 or S_2?
By inspection, we can clearly see that S_2 is larger than S_1.

You should be able to see this by inspection, but if you need "proof," here are the respective sums:

$S_1 = 1 + 12 + 123 + 1,234 + 12,345 + 123,456 + 1,234,567 + 12,345,678 + 123,456,789$
$\quad = 137,174,205$

and
$$S_2 = 987{,}654{,}321 + 87{,}654{,}321 + 7{,}654{,}321 + 654{,}321 + 54{,}321 + 4{,}321 + 321 + 21 + 1$$
$$= 1{,}083{,}676{,}269$$

This time we will complicate the problem a bit by adding zeros to make all of the numbers of S_1 nine digits long. Now, which is larger, S_3 or S_2 ?
$$S_3 = 100{,}000{,}000 + 120{,}000{,}000 + 123{,}000{,}000 + 123{,}400{,}000 + 123{,}450{,}000 +$$
$$123{,}456{,}000 + 123{,}456{,}700 + 123{,}456{,}780 + 123{,}456{,}789$$
$$S_2 = 987{,}654{,}321 + 87{,}654{,}321 + 7{,}654{,}321 + 654{,}321 + 54{,}321 + 4{,}321 + 321 + 21 + 1$$

The amazing thing is that both sums are equal!
That is, both series have a sum of $S_2 = S_3 = 1{,}083{,}676{,}269$.

The proof of this amazing result can be obtained by cleverly writing the numbers to be added rather than simply adding them in a traditional manner. You will notice that by writing them in the two columns as in figure 2-3, and organizing them as we have them, the result is obvious.

S_2	S_3
9 8 7 6 5 4 3 2 1	1 2 3 4 5 6 7 8 9
8 7 6 5 4 3 2 1	1 2 3 4 5 6 7 8 0
7 6 5 4 3 2 1	1 2 3 4 5 6 7 0 0
6 5 4 3 2 1	1 2 3 4 5 6 0 0 0
5 4 3 2 1	1 2 3 4 5 0 0 0 0
4 3 2 1	1 2 3 4 0 0 0 0 0
3 2 1	1 2 3 0 0 0 0 0 0
2 1	1 2 0 0 0 0 0 0 0
1	1 0 0 0 0 0 0 0 0

Figure 2-3

Notice that each (vertical) column of digits—the units, tens, hundreds, and so on—for both S_2 and S_3, have the same sum, which is summarized in figure 2-4.

S_2	S_3
$9 \cdot 1 = 9$	$1 \cdot 9 = 9$
$8 \cdot 2 = 16$	$2 \cdot 8 = 16$
$7 \cdot 3 = 21$	$3 \cdot 7 = 21$
$6 \cdot 4 = 24$	$4 \cdot 6 = 24$
$5 \cdot 5 = 25$	$5 \cdot 5 = 25$
$4 \cdot 6 = 24$	$6 \cdot 4 = 24$
$3 \cdot 7 = 21$	$7 \cdot 3 = 21$
$2 \cdot 8 = 16$	$8 \cdot 2 = 16$
$1 \cdot 9 = 9$	$9 \cdot 1 = 9$

Figure 2-4

That is, the sums of the powers of ten places of each sum are:

Units = 9

Tens = 16

Hundreds = 21

Thousands = 24, and so on.

And therefore, $S_2 = S_3$.

Again, you have a more clever way of determining a sum than merely adding numbers as would be indicated.

CAN YOU LOCATE THE NUMBER?

Suppose we write the natural numbers in the arrangement of the triangle as in figure 2-5.

```
                       1
                    2  3  4
                 5  6  7  8  9
             10 11 12 13 14 15 16
          17 18 19 20 21 22 23 24 25
       26 27 28 ...
```

Figure 2-5

In which row would we find the number 2,000? In which column would 2,000 be located?

We should first ask: Do we have enough information to answer the question? Will we actually have to enlarge the triangular arrangement of numbers to the point where 2,000 appears? Here is a situation that will circumvent actual counting by looking for the number patterns.

You will notice that each row ends in a square number—one that corresponds to the number of the row. That is, the third row, for example, ends in 3^2, or 9. The fifth row ends in 25, which is 5^2; and so the nth row would end in n^2.

The first number in each row is one greater than the last number of the previous row, which is the square of the number of that row. So, for the nth row, we can write this first number as $(n-1)^2 + 1$.

The middle number of each row is the average of the first and the last numbers of that row, since the numbers are consecutive. Hence, for the nth row, the middle term is $\frac{[(n-1)^2+1]+n^2}{2} = \frac{n^2-2n+1+1+n^2}{2} = n^2 - n + 1$.

Now to our quest to locate the number 2,000. Since $44^2 = 1,936$, and $45^2 = 2,025$, the number 2,000 must be between 44^2 and 45^2. Therefore, it must be in the row that ends in 45^2, which means it must lie in row 45.

Using our formula for the first term of a row, $(n-1)^2 +1$, we get the first term of row 45 to be: $(45-1)^2 +1=1,937$. The last term is $45^2 = 2,025$, and the middle number is the average of these two, 1,981. We can then easily determine that 2,000 is in the twentieth column.

YOUR FAVORITE DIGIT

You might try this little calculating "trick" with a friend. Write the following number on a piece of paper: 12,345,679 (notice the 8 is missing). Then have your friend multiply his favorite digit (1–9) by 9 times this number. To your friend's amazement, the product will consist of only his favorite digit.

For example, consider the following three cases (figure 2-6), where your friend may select as favorite digits 4, 2, and 5.

Favorite digit **4**	*Favorite digit* **2**	*Favorite digit* **5**
Multiplication $4\cdot 9 = 36$	Multiplication of $2\cdot 9 = 18$	Multiplication of $5\cdot 9 = 45$
$12,345,679\cdot 36 = 444,444,444$	$12,345,679\cdot 18 = 222,222,222$	$12,345,679\cdot 45 = 555,555,555$

Figure 2-6

From this you can see how to have some fun with arithmetic.

A CUTE LITTLE CALCULATING SCHEME

Select any three-digit number and write it twice to form a six-digit number. For example, if you choose the number 357, then write the six-digit number 357,357. We now divide this number by 7, then divide the resulting quotient by 11, and then divide that quotient by 13:

$$\frac{357,357}{7} = 51,051$$

$$\frac{51,051}{11} = 4,641$$

$$\frac{4,641}{13} = 357$$

This is the first number you started with!

The reason that this works as it does is that to form the original six-digit number, you actually multiplied the original three-digit number by 1,001. That is, $357 \cdot 1,001 = 357,357$.

But $1,001 = 7 \cdot 11 \cdot 13$. Therefore, by dividing successively by 7, 11, and 13, we have undone the original multiplication by 1,001, leaving the original number.

While we are on the number 1,001, we can see how this number can also help us multiply other number combinations.

$$221 \cdot 77 = (17 \cdot 13)(11 \cdot 7) = 1,001 \cdot 17 = 1,000 \cdot 17 + 1 \cdot 17 = 17,017$$
$$264 \cdot 91 = (24 \cdot 11)(13 \cdot 7) = 1,001 \cdot 24 = 1,000 \cdot 24 + 1 \cdot 24 = 24,024$$
$$407 \cdot 273 = (37 \cdot 11)(3 \cdot 7 \cdot 13) = 1,001 \cdot 111 = 1,000 \cdot 111 + 1 \cdot 111 = 111,111$$

This last number, 111,111, can be gradually expanded as:

$$111 \quad = \quad 3 \cdot 37$$
$$1,111 \quad = \quad 11 \cdot 101$$
$$11,111 \quad = \quad 41 \cdot 271$$
$$111,111 \quad = \quad 3 \cdot 7 \cdot 11 \cdot 13 \cdot 37$$

Therefore, $\dfrac{111,111}{37} = 3,003$, $\dfrac{111,111}{143} = \dfrac{111,111}{11 \cdot 13} = 777$, and so on.

So, for example, suppose you want to multiply $74 \cdot 3,003$.

$$74 \cdot 3,003 = (2 \cdot 37) \cdot 3,003 = (2 \cdot 37) \cdot \frac{111,111}{37} = 2 \cdot 111,111 = 222,222$$
$$858 \cdot 777 = (2 \cdot 3 \cdot 11 \cdot 13) \cdot 777 = (2 \cdot 3 \cdot 11 \cdot 13) \cdot \frac{111,111}{11 \cdot 13} = 2 \cdot 3 \cdot 111,111 = 666,666$$

Yes, arithmetic can expose some hidden numerical treasures.

THE 777 NUMBER TRICK

As you have seen before and will continue to see in this chapter, we can get a nice appreciation for numbers by using arithmetic in the context of a recreational setting. You might want to try this with a friend. Have your friend select any number between 500 and 1,000; then have him add 777 to this number. Because the sum exceeds 1,000, we then have him remove the thousands digit and add it to the units digit of the sum. Now have him subtract the two numbers—this sum and the one he originally selected. You can tell him that he must now have arrived at 222.

Let's try one together now. Suppose the friend selects the number 600. He then adds to it 777, to get: $600 + 777 = 1,377$. He now removes the 1 and adds it to the units digit to get 378. Subtracting these two numbers, $600 - 378 = 222$.

You may wonder why this works as it does. For every selected number between 500 and 1,000 you will always get a 1 in the thousands place when this number is added to 777. Dropping the 1 and adding it to the units digit is tantamount to merely subtracting 999 from the number. That is, $-999 = -1,000 + 1$.

If we now represent the selected number as n, then what is being done is:

$$n - (n + 777 - 999) = n - n - 777 + 999 = 222$$

Remember, n represents the number randomly selected.

Suppose we would have used a number other than 777 as our "magic" number, say, 591, then we would have our friend end up with 408 every time, regardless of which number he chose between 500 and 1,000. For a "magic" number of 733, the end result will always be 266. Remember, the selected number cannot be less than 500 or else you might not get a sum in the thousands, and at the same time the selected number should not be greater than 999 or you might get a 2 in the thousands place, which would ruin this scheme.

INDIRECT CALCULATION

There are times when simple arithmetic is not enough to answer a question. These are times when logical reasoning must be used to buttress the arithmetic. A case in point can be seen from this neat little problem.

> A woman with her three daughters passes her neighbor's house and he asks her how old her three daughters are. She responds that coincidentally the product of their ages is 36 and the sum of their ages is the same number as his address. He looks puzzled as he stares at the house number, finding no solution, and then gets even more puzzled when the woman tells him that she almost forgot one essential piece of information: her oldest daughter's name is Lisa. This really baffles him. How can this man determine the ages of the woman's daughters? (We are only dealing with integer ages.)

To determine the ages of the three daughters, the man makes a chart of the possible ages (figure 2-7), that is, the three numbers whose product is 36:

Age of daughter a	Age of daughter b	Age of daughter c	Age sum $a + b + c$
1	1	36	38
1	2	18	21
1	3	12	16
1	4	9	14
1	6	6	13
2	2	9	13
2	3	6	11
3	3	4	10

Figure 2-7

From the chart in figure 2-7, we see that with all the possible products of 36, there is only one case where the man would need additional information to figure out the answer; that is, if the sum of their ages is 13. So, when the woman tells the man that she left out an essential piece of information, it must have been to differentiate between the two sums of 13. When she mentioned that her oldest daughter is named Lisa, that indicated that there was only one older daughter, thus eliminating the possibility of twins of age 6. Hence, the ages of the three daughters are 2, 2, and 9.

Here arithmetic calculation alone did not help us answer the question.

A COMMON TRICK ON THE INTERNET

This interactive trick has been circulating about the Internet in various forms. Most users are baffled that the computer can respond by determining a secretly selected number—when clearly it cannot see it nor was it revealed in any obvious way.

The interaction between the computer Web site and the user might go like this (figure 2-8):

The computer	The player
Write on paper any five-digit number (not one with five of the same digits).	35,630
Rearrange the digits of your selected number.	53,306
Subtract these two numbers.	$53,306 - 35,630 = 17,767$
Delete any digit (1–9) of this difference, *but not* 0, and keep it secret.	17,6~~7~~6
Rearrange the digits of this number and input it in the computer.	7, 6, 1, 6
I can tell you which digit you removed. It was a 7.	Mystified!

Figure 2-8

The user tends to try this a few times only to find that the computer can somehow determine which number he has removed. How is this possible? Yes, some knowledge of number properties will help you figure out this secret.

The important fact to know here is that the difference of two numbers with the same digits will always have a digit sum that is a multiple of 9—that is, 9, 18, 27, 36, 45, and so on. So when you tell the computer which digits remain after you have deleted one digit, the computer simply takes the sum of the digits and selects the digit that would bring that digit sum to the next multiple of 9. That must be the digit you removed earlier. So you can see that this "trick" is completely dependent on the notion that the difference of two numbers with the same digit sum must have a digit sum which is a multiple of 9. Such arithmetic trivia gives one a better grasp of number properties.

NUMBER TRICKS WITH THE NUMBER 9

While we are on the properties of 9, we can construct many entertaining number tricks that involve the number 9, which results from the fact that 9 is one less than the base of our number system, 10. Many of these tricks rely on the fact that any number which is a multiple of 9 will have a digit sum that is a multiple of 9, and that reduces, through continuous digit sums taken, to the number 9 itself.

For example, let's make up just such a number trick. Tell your friend to select any number and add his age to it. Then have him add to this result the last two digits of his telephone number and then multiply this number by 18. He then takes the sum of the digits of his last number and continues to take the digit sum until he has a single digit, and you will impress him by telling him that this last digit is a 9.

Let's try this with the selection of the number 39 and add our age (37) to it to get 76. We then add the last two digits of our telephone number (31) to get 107 and then multiply this number by 18 to get 1,926. The digit sum is $1 + 9 + 2 + 6 = 18$, whose digit sum is $1 + 8 = 9$.

This works because when we multiplied by 18, we made sure that the final result would be a multiple of 9, which eventually always yields a digit sum of 9.

SOME MULTIPLICATION TRICKS

Most people know that to multiply by 10 we merely have to tag a zero onto the number being multiplied by 10. For example, when we multiply 78 by 10 we get 780.

The simpler a mathematical "trick" is, the more attractive it tends to be. Here is a very nifty way to multiply by 11. This one always gets a rise out of the unsuspecting mathematics-phobic person, because it is so simple that it is even easier than doing it on a calculator!

The rule is very simple:

**To multiply a two-digit number by 11, just add the two digits
and place this sum between the two digits.**

Let's try using this technique. Suppose you wish to multiply 45 by 11. According to the rule, add 4 and 5 and place their sum, 9, between the 4 and 5 to get 495.

This can get a bit more difficult if the digit sum is a two-digit number. What do we do in that case? We no longer have a single digit to place between the two original digits. So if the sum of the two digits is greater than 9, then we place the units digit between the two digits of the number being multiplied by 11 and "carry" the tens digit to be added to the hundreds digit of the multiplicand.[3] Let's try it with $78 \cdot 11$. Here $7 + 8 = 15$. We place the 5 between the 7 and 8, and add the 1 to the 7, to get $[7 + 1][5][8]$ or 858.

You may legitimately ask if the rule also holds when 11 is multiplied by a number of more than two digits.

Let's go right for a larger number such as 12,345 and multiply it by 11. Here we begin at the units digit and add every pair of digits going to the left.

$1[1 + 2][2 + 3][3 + 4][4 + 5]5 = 135,795.$

If the sum of two digits is greater than 9, then use the procedure described before: place the units digit of that sum appropriately and carry the tens digit. We will do one of these here.

Multiply 456,789 by 11. We carry the process step by step:

$$4[4 + 5][5 + 6][6 + 7][7 + 8][8 + 9]9$$
$$4[4 + 5][5 + 6][6 + 7][7 + 8][17]9$$
$$4[4 + 5][5 + 6][6 + 7][7 + 8 + 1][7]9$$
$$4[4 + 5][5 + 6][6 + 7][16][7]9$$
$$4[4 + 5][5 + 6][6 + 7 + 1][6][7]9$$
$$4[4 + 5][5 + 6][14][6][7]9$$
$$4[4 + 5][5 + 6 + 1][4][6][7]9$$
$$4[4 + 5][12][4][6][7]9$$
$$4[4 + 5 + 1][2][4][6][7]9$$
$$4[10][2][4][6][7]9$$
$$[4 + 1][0][2][4][6][7]9$$
$$[5][0][2][4][6][7]9$$
$$5,024,679$$

This rule for multiplying by 11 ought to be shared with your friends. Not only will they be impressed with your cleverness, but they may also appreciate knowing this shortcut.

---

3. The *multiplicand* is the number that is multiplied by another number, the *multiplier*. In arithmetic, the multiplicand and the multiplier are interchangeable, depending on how the problem is stated, because the result is the same if the two are reversed—for example, $2 \cdot 3$ and $3 \cdot 2$. Therefore, $2 \cdot 3$ means "add 2 three times," whereas $3 \cdot 2$ means "add 3 two times."

There are quick multiplication tricks for special pairs of numbers. For example, if we are multiplying two numbers that are the same except for the units digit, which have a sum of 10, we reduce one number to the next lower multiple of 10 and increase the other to the next higher multiple of 10 and multiply these numbers. Then, we add to this product the product of the unit digits of the original two numbers. The result is our desired product.

Here are a few examples of this procedure:

$$
\begin{aligned}
13 \cdot 17 &= 10 \cdot 20 + 3 \cdot 7 &&= 200 + 21 &&= 221 \\
36 \cdot 34 &= 30 \cdot 40 + 6 \cdot 4 &&= 1{,}200 + 24 &&= 1{,}224 \\
64 \cdot 66 &= 60 \cdot 70 + 4 \cdot 6 &&= 4{,}200 + 24 &&= 4{,}224 \\
72 \cdot 78 &= 70 \cdot 80 + 2 \cdot 8 &&= 5{,}600 + 16 &&= 5{,}616 \\
104 \cdot 106 &= 100 \cdot 110 + 4 \cdot 6 &&= 11{,}000 + 24 &&= 11{,}024 \\
307 \cdot 303 &= 300 \cdot 310 + 7 \cdot 3 &&= 93{,}000 + 21 &&= 93{,}021
\end{aligned}
$$

SOME AIDS IN MENTAL CALCULATION

There are many methods for multiplying two-digit numbers up to 20. Here are some of these methods, which might also provide some insight to others.

Multiply $18 \cdot 17$ mentally. Here we seek to extract a multiple of 10 first.

$18 \cdot 17 = 18 \cdot 10 + 18 \cdot 7 = 18 \cdot 10 + 10 \cdot 7 + 8 \cdot 7 = 180 + 70 + 56 = 306$
or
$18 \cdot 17 = 10 \cdot 17 + 8 \cdot 17 = 170 + 136 = 306$

Another method of multiplication would seek to get more familiar factors:

$18 \cdot 17 = (20 - 2) \cdot 17 = 20 \cdot 17 - 2 \cdot 17 = 340 - 34 = 306$
or
$18 \cdot 17 = 18 \cdot (20 - 3) = 18 \cdot 20 - 18 \cdot 3 = 360 - 54 = 306$

Here is an entirely different method to multiply two such numbers:

Step 1: Add one number to the units digit of the other number: $18 + 7 = 25$
Step 2: Tag a zero onto the number: 250
Step 3: Multiply the two units digits: $8 \cdot 7 = 56$
Step 4: Add the results of steps 2 and 3: $250 + 56 = 306$.

Try this technique with other two-digit numbers up to 20.

A QUICK METHOD TO MULTIPLY BY DIVISORS OF POWERS OF 10

We all know that multiplying by powers of 10 is relatively easy. You need only tag the relevant number of zeros onto the number being multiplied by the power of ten. That is, 685 times 1,000 is 685,000.

However, multiplying by factors (or divisors) of powers of 10 is just a bit more involved but, in many cases, can also be done mentally. Let's consider multiplying by 25 (a factor of 100).

Since $25 = \dfrac{100}{4}$, $16 \cdot 25 = 16 \cdot \dfrac{100}{4} = \dfrac{16}{4} \cdot 100 = 4 \cdot 100 = 400$.

Here are some more examples of this:

$$38 \cdot 25 = 38 \cdot \frac{100}{4} = \frac{38}{4} \cdot 100 = \frac{19}{2} \cdot 100 = 9.5 \cdot 100 = 950$$

$$1.7 \cdot 25 = \frac{17}{10} \cdot \frac{100}{4} = \frac{17}{4} \cdot \frac{100}{10} = 4.25 \cdot 10 = 42.5$$

(Remember that $\dfrac{16}{4} = 4$ and $\dfrac{1}{4} = .25$; therefore, $\dfrac{17}{4} = \dfrac{16}{4} + \dfrac{1}{4} = 4 + 0.25 = 4.25$.)

In an analogous fashion, we can multiply by 125, since $125 = \dfrac{1,000}{8}$.

Here are some examples of multiplication by 125—mentally!

$$32 \cdot 125 = 32 \cdot \frac{1,000}{8} = \frac{32}{8} \cdot 1,000 = 4 \cdot 1,000 = 4,000$$

$$78 \cdot 125 = 78 \cdot \frac{1,000}{8} = \frac{78}{8} \cdot 1,000 = \frac{39}{4} \cdot 1,000 = 9.75 \cdot 1,000 = 9,750$$

$$3.4 \cdot 125 = 3.4 \cdot \frac{1,000}{8} = \frac{3.4}{8} \cdot 1,000 = \frac{1.7}{4} \cdot 1,000 = 0.425 \cdot 1,000 = 425$$

When multiplying by 50, you can use $50 = \dfrac{100}{2}$, or when multiplying by 20, you can use $20 = \dfrac{100}{5}$.

Practice with these special numbers will clearly be helpful to you, since you will be able to do many calculations faster than the time it takes to find and then turn on your calculator!

MENTALLY SQUARING NUMBERS ENDING IN 5

Suppose we want to square 45. That is, $45^2 = 45 \cdot 45 = 2,025$.

The process requires three steps:

Step 1: Multiply the multiples of 10 higher and lower than the number to be squared
$40 \cdot 50 = 2,000$

Step 2: Square the units digit 5: $5 \cdot 5 = 25$

Step 3: Add the two results from steps 1 and 2: $2,000 + 25 = 2,025$

In case you are curious why this works, take a look at the following:

$$
\begin{aligned}
45 \cdot 45 &= (40 + 5) \cdot (50 - 5) = 40 \cdot 50 + 40 \cdot (-5) + 5 \cdot 50 + 5 \cdot (-5) \\
&= 40 \cdot 50 - 40 \cdot 5 + 5 \cdot 50 - 5 \cdot 5 \\
&= 40 \cdot 50 + 5 \cdot (-40 + 50 - 5) \\
&= \mathbf{40 \cdot 50 + 5 \cdot 5} \\
&= 2,000 + 25 \\
&= 2,025
\end{aligned}
$$

As a treat for the purists, we offer a short proof of this by using elementary algebra.
We know that $(u + v)^2 = u^2 + 2uv + v^2$.
We let x be the multiple of 5 to be squared.

$x^2 = (10a + 5)^2 = 100a^2 + 100a + 25 = 100a(a + 1) + 25 = a(a + 1) \cdot 100 + 25$

or

$x^2 = a \cdot 10 \cdot (a + 1) \cdot 10 + 25$, which says, algebraically, just what we did in the three steps above

Here is an example of how to interpret this:
Consider again $45 \cdot 45$:

Here $a = 4$; then $x^2 = a \cdot 10 \cdot (a + 1) \cdot 10 + 25 = 4 \cdot 10 \cdot (4 + 1) \cdot 10 + 25 = \mathbf{40 \cdot 50 + 5 \cdot 5}$
$= 2,025$

Another illustration:
For the multiplication $175 \cdot 175$:[4]

$a = 17$; $x^2 = a \cdot 10 \cdot (a + 1) \cdot 10 + 25 = 17 \cdot 10 \cdot (17 + 1) \cdot 10 + 25 = \mathbf{170 \cdot 180 + 5 \cdot 5}$
$= 30,600 + 25 = 30,625$

---

4. Here we can use the "trick" for $18 \cdot 17$ to get $17 \cdot 18 = 306$.

Another way of looking at this technique is by considering the pattern that evolves below. Look at the list below and see if you can identify the pattern of the two-digit numbers squared.

$05^2 =$ **25**
$15^2 =$ **225**
$25^2 =$ **625**
$35^2 = 1,$**225**
$45^2 = 2,0$**25**
$55^2 = 3,0$**25**
$65^2 = 4,2$**25**
$75^2 = 5,6$**25**
$85^2 = 7,22$**5**
$95^2 = 9,0$**25**

You will notice that in each case the square ends with 25 and the preceding digits are determined as follows:

$05^2 = \mathbf{0,0}25, \quad 0 = 0 \cdot 1$
$15^2 = \mathbf{0,2}25, \quad 2 = 1 \cdot 2$
$25^2 = \mathbf{0,6}25, \quad 6 = 2 \cdot 3$
$35^2 = \mathbf{1,2}25, \quad 12 = 3 \cdot 4$
$45^2 = \mathbf{2,0}25, \quad 20 = 4 \cdot 5$
$55^2 = \mathbf{3,0}25, \quad 30 = 5 \cdot 6$
$65^2 = \mathbf{4,2}25, \quad 42 = 6 \cdot 7$
$75^2 = \mathbf{5,6}25, \quad 56 = 7 \cdot 8$
$85^2 = \mathbf{7,2}25, \quad 72 = 8 \cdot 9$
$95^2 = \mathbf{9,0}25, \quad 90 = 9 \cdot 10$

The same rule can be extended to three-digit numbers and beyond. Take, for example, $235^2 = \mathbf{55,2}25$; $552 = 23 \cdot 24$. The advantage of mental arithmetic tends to lose its attractiveness when we exceed two-digit numbers since we must multiply two two-digit numbers—something usually not easily done mentally.

A MORE GENERAL METHOD TO DO MENTAL MULTIPLICATION

In the previous cases, we used the properties of binomial multiplication. Here we will use this in a more general way. You will recall the following binomial multiplication:

$(u + v)(u - v) = u^2 - uv + uv - v^2 = u^2 - v^2$, where u and v can take on any values that would be convenient to us.

We can apply this to the multiplication of $93 \cdot 87$. We notice that the two numbers are symmetrically distanced from 90. This leads us to the following:

$93 \cdot 87 = (90 + 3)(90 - 3) = 90^2 - 3^2 = 8,100 - 9 = 8,091$

To review what we did:

$$\begin{array}{lll} \text{Step 1: } 90^2 & = 8,100 \\ \text{Step 2: } 3^2 & = 9 \\ \text{Step 3: subtract} & = 8,091 \end{array}$$

Here are a few further examples—then you ought to practice on your own to bring this into your arsenal of arithmetic skills.

$$42 \cdot 38 = (40 + 2)(40 - 2) = 40^2 - 2^2 = 1,600 - 4 = 1,596$$
$$21 \cdot 19 = (20 + 1)(20 - 1) = 20^2 - 1^2 = 400 - 1 = 399$$
$$64 \cdot 56 = (60 + 4)(60 - 4) = 60^2 - 4^2 = 3,600 - 16 = 3,584$$

In order to really become comfortable with these mental multiplication methods, we offer you a few examples to guide you along:

$$67 \cdot 63 = (65 + 2)(65 - 2) = 65^2 - 2^2 = [60 \cdot 70 + 5 \cdot 5] - 4 = 4,225 - 4 = 4,221$$
$$26 \cdot 24 = (25 + 1)(25 - 1) = 25^2 - 1^2 = [20 \cdot 30 + 5 \cdot 5] - 1 = 625 - 1 = 624$$
$$138 \cdot 132 = (135 + 3)(135 - 3) = 135^2 - 3^2 = [130 \cdot 140 + 5 \cdot 5] - 9$$
$$= [(13 \cdot 14) \cdot 100 + 5 \cdot 5] - 9 = [(170 + 12) \cdot 100 + 5 \cdot 5] - 9$$
$$= 182 \cdot 100 + 5 \cdot 5 - 9 = 18,200 + 25 - 9 = 18,200 + 16 = 18,216$$

MENTAL ARITHMETIC CAN BE MORE CHALLENGING— BUT USEFUL!

On paper you might think that some of the methods we are presenting as mental arithmetic are more complicated than doing the work with the traditional algorithms. Yet, with practice, some of these methods for two-digit numbers will become simpler to do mentally than writing them on paper and using the traditional algorithms.

Take, for example, the multiplication $95 \cdot 97$. The following steps can be done mentally (with a modicum of practice, naturally!):

Step 1: $95 + 97$ = 192

Step 2: Delete the hundreds digit = 92

Step 3: Tag two zeros onto the number = 9,200

Step 4: $(100 - 95)(100 - 97) = 5 \cdot 3$ = 15

Step 5: Add the last two numbers = 9,215

Here is another example of this technique:

$93 \cdot 96 = \ldots$

$93 + 96 = 189$

$\overline{1}89$ (delete the hundreds digit)

Tag on two zeros = 8,900

Then add $(100 - 93)(100 - 96) = 7 \cdot 4 = 28$ to get 8,928

This technique also works when seeking the product of two numbers that are further apart:

$89 \cdot 73 = \ldots$

$89 + 73 = 162$

$\overline{1}62$ (delete the hundreds digit)

Tag on two zeros = 6,200

Then add $(100 - 89)(100 - 73) = 11 \cdot 27 = 297$ to get 6,497

For those who might be curious why this technique works, we can show you the simple algebra that will justify it.

We begin with the two-digit numbers $100 - a$ and $100 - b$ (where $0 < a, b < 100$).

Step 1: $(100 - a) + (100 - b) = 200 - a - b$

Step 2: Delete the hundreds digit—which means subtracting 100 from the number: $(200 - a - b) - 100 = 100 - a - b$

Step 3: Tag on two zeros, which means multiply by 100: $(100 - a - b) \cdot 100 = 10,000 - 100a - 100b$

Step 4: $a \cdot b$

Step 5: Add the last two results:
$$10,000 - 100a - 100b + a \cdot b$$
$$= 100(100 - a) - (100b - ab)$$
$$= 100(100 - a) - b(100 - a) = (100 - a)(100 - b)$$

which is what we set out to show. Now you just need to practice this method to master it!

CHALLENGING YOUR FRIEND TO A SPEED ADDITION CONTEST

Now that we have considered some alternative methods for doing common arithmetic, it might be nice to use number relationships to play a little trick on your friends. Remember, for every "trick" in mathematics there is some fine relationship or mathematical property that allows us to do it. This can be quite instructive. So now let's show off our ability to add numbers faster than any of our friends.

We will write the numbers to be added in a vertical column. To begin, have a friend randomly select a five-digit number. Let's suppose our friend selected the number **45,712**.

Now have him select another five-digit number, and write it beneath the first number.
45,712
31,788

Now it is our turn to select a five-digit number to add to the first two numbers:
45,712
31,788
68,211

The next five-digit number is selected by our friend:
45,712
31,788
68,211
41,527

We will select the next five-digit number:
45,712
31,788
68,211
41,527
58,472

We can now simply write the sum below:
45,712
31,788
68,211
41,527
+ 58,472
245,710

Clearly we will have arrived at the sum before our friend. But how did we do this, asks the friend?

Before we expose the secret of this addition problem, take a look at the following example:

(Friend selects)	12,915
(Friend selects)	12,708
(We select)	87,291
(Friend selects)	31,535
(We select)	+ 68,464
	212,913

When we inspect the second and third numbers, we notice that the sum is 99,999.

"Surprisingly," the same is true for the last two numbers. Remember, we selected one number for each of these pairs of numbers, intentionally making them sum to 99,999. The remainder of the addition is then easy: $99,999 + 99,999 = 100,000 - 1 + 100,000 - 1 = 200,000 - 2$.

Therefore to get the entire sum, we simply add 200,000 + the starting number – 2. For our example, $200,000 + 12,915 - 2 = 212,913$.

Here is another example (figure 2-9) to help you master this trick:

$$
\begin{array}{rcl}
45,712 & & 45,712 \\
\left.\begin{array}{l} 31,788 \\ 68,211 \end{array}\right\} & & 99,999 \\
+\left.\begin{array}{l} 41,527 \\ 58,472 \end{array}\right\} & + & 99,999 \\
\hline
245,710 & & 245,710 \\
 & = & \\
 & & 200,000 \\
 & - & 2 \\
 & + & 45,712 \\
\hline
 & & 245,710
\end{array}
$$

Figure 2-9

We could eventually determine the sum after the first number your friend offers.

THE ADDITION TRICK OF SIX NUMBERS

This trick is fun and, as before, will show how we can analyze a seemingly baffling result through simple algebra. We begin by asking our trusty friend to select any three-digit number with no two like digits (omitting zero). Then, have him make five other numbers with these same digits.[5] Suppose the friend selected the number 473, then the list of all numbers formed from these digits is:

473

437

347

374

743

734

We can get the sum of these numbers, 3,108, faster than he can even write the numbers. How can this be done so fast? All we actually have to do is get the sum of the digits of the original number (here: we get $4 + 7 + 3 = 14$) and then multiply 14 by 222 to get 3,108, which is the required sum. Why 222? Let us inspect some of the many quirks of number properties using simple algebra:

Consider the number $\overline{abc} = 100a + 10b + c$, where a, b, c can be any of the digits 1, 2, 3, ... , 9.

The sum of the digits is $a + b + c$. We now represent all of the six numbers of these digits on our list:

$$100a \quad + 10b \quad + c$$
$$100a \quad + 10c \quad + b$$
$$100b \quad + 10a \quad + c$$
$$100b \quad + 10c \quad + a$$
$$100c \quad + 10a \quad + b$$
$$100c \quad + 10b \quad + a$$

This equals:

$$100(2a + 2b + 2c) + 10(2a + 2b + 2c) + 1(2a + 2b + 2c)$$
$$= 200(a + b + c) + 20(a + b + c) + 2(a + b + c)$$
$$= 222(a + b + c)$$

which is 222 times the sum of the digits. If you really want to be slick and beat your friend dramatically, then you might want to have the following (figure 2- 10) on a small piece of paper for easy reference:

---

5. If you have been exposed to permutations before, you will recognize that these six numbers are the *only* ways to write a number with three different digits—which is why we ask for three *distinct* digits in the first place!

Digit sums	6	7	8	9	10	11	12	13	14	15
Six-number sum	1,332	1,554	1,776	1,998	2,220	2,442	2,664	2,886	3,108	3,330

Digit sum	16	17	18	19	20	21	22	23	24
Six-number sum	3,552	3,774	3,996	4,218	4,440	4,662	4,884	5,106	5,328

Figure 2-10

So that's all we need to avoid the actual addition.

A NUMBER TRICK BASED ON A FIBONACCI PROPERTY

You might like to experience a cute little mathematical trick that is based on the famous Fibonacci numbers.[6] It will give you a chance to impress others with your ability to add ten numbers mentally. Ask your friend to select any two natural numbers and add them. Then he is to add the newly found sum to the second of the two original numbers. Again, have him add the sum to the last of the previous number list. This should be continued until ten numbers are listed. Here is an example of what you are to have your friend do: suppose z_1 and z_2 are the first two numbers selected by your friend. Then the third number is z_3 (= $z_1 + z_2$). Then he is to add z_2 and z_3 to get their sum, z_4. For convenience we will list these vertically.

z_1

z_2

z_3

$z_4 (= z_2 + z_3)$

Continue along until there are ten numbers in the list: $z_1, z_2, z_3, z_4, z_5, z_6, z_7, z_8, z_9, z_{10}$

A quick glance at these numbers—and recalling your newly acquired skill at multiplying by 11 mentally (pages 66–67)—should allow you to get the sum immediately. So how is this actually done? We simply mutiply the seventh number, z_7, by 11, and we have the

6. A. S. Posamentier and I. Lehmann, *The Fabulous Fibonacci Numbers* (Amherst, NY: Prometheus Books, 2007), pp. 142–43.

sum of all ten listed numbers. Before we examine this more closely, let's try an actual example of this trick. We will begin by letting the first two numbers be 123 and 456.

Then we have:

z_1						$=$	123
z_2						$=$	456
z_3	$=$	$z_1 + z_2$	$=$	123	$+$	456	$=$ 579
z_4	$=$	$z_2 + z_3$	$=$	456	$+$	579	$=$ $1{,}035$
z_5	$=$	$z_3 + z_4$	$=$	579	$+$	$1{,}035$	$=$ $1{,}614$
z_6	$=$	$z_4 + z_5$	$=$	$1{,}035$	$+$	$1{,}614$	$=$ $2{,}649$
z_7	$=$	$z_5 + z_6$	$=$	$1{,}614$	$+$	$2{,}649$	$=$ $4{,}263$
z_8	$=$	$z_6 + z_7$	$=$	$2{,}649$	$+$	$4{,}263$	$=$ $6{,}912$
z_9	$=$	$z_7 + z_8$	$=$	$4{,}263$	$+$	$6{,}912$	$=$ $11{,}175$
z_{10}	$=$	$z_8 + z_9$	$=$	$6{,}912$	$+$	$11{,}175$	$=$ $18{,}087$

$$s = z_1 + z_2 + z_3 + z_4 + z_5 + z_6 + z_7 + z_8 + z_9 + z_{10}$$
$$= 123 + 456 + 579 + 1{,}035 + 1{,}614 + 2{,}649 + 4{,}263 + 6{,}912 + 11{,}175 + 18{,}087$$
$$= 46{,}893$$

The trick is to see that the sum will always be equal to 11 times the seventh number: $46{,}893 = 11 \cdot 4{,}263$.

That is, $s = z_1 + z_2 + z_3 + z_4 + z_5 + z_6 + z_7 + z_8 + z_9 + z_{10} = 11 \cdot z_7$.

To appreciate why this works, we will inspect the process algebraically:

z_1							$=$	z_1
z_2							$=$	z_2
z_3	$=$	$z_1 + z_2$					$=$	$z_1 + z_2$
z_4	$=$	$z_2 + z_3$	$=$	z_2	$+$	$(z_1 + z_2)$	$=$	$z_1 + 2z_2$
z_5	$=$	$z_3 + z_4$	$=$	$(z_1 + z_2)$	$+$	$(z_1 + 2z_2)$	$=$	$2z_1 + 3z_2$
z_6	$=$	$z_4 + z_5$	$=$	$(z_1 + 2z_2)$	$+$	$(2z_1 + 3z_2)$	$=$	$3z_1 + 5z_2$
z_7	$=$	$z_5 + z_6$	$=$	$(2z_1 + 3z_2)$	$+$	$(3z_1 + 5z_2)$	$=$	$5z_1 + 8z_2$
z_8	$=$	$z_6 + z_7$	$=$	$(3z_1 + 5z_2)$	$+$	$(5z_1 + 8z_2)$	$=$	$8z_1 + 13z_2$
z_9	$=$	$z_7 + z_8$	$=$	$(5z_1 + 8z_2)$	$+$	$(8z_1 + 13z_2)$	$=$	$13z_1 + 21z_2$
z_{10}	$=$	$z_8 + z_9$	$=$	$(8z_1 + 13z_2)$	$+$	$(13z_1 + 21z_2)$	$=$	$21z_1 + 34z_2$

By adding these ten numbers, we get:

$$s = z_1 + z_2 + z_3 + z_4 + z_5 + z_6 + z_7 + z_8 + z_9 + z_{10}$$
$$= 55z_1 + 88z_2$$
$$= 11 \cdot (5z_1 + 8z_2)$$
$$= 11 \cdot (z_5 + z_6)$$
$$= 11 \cdot z_7$$

By inspecting the coefficients in the above list, you may spot the Fibonacci numbers: 1, 1, 2, 3, 5, 8, 13, 21, and 34. These ubiquitous numbers are worth further investigation, and we offer you a resource: *The Fabulous Fibonacci Numbers* (by Alfred S. Posamentier and Ingmar Lehmann, [Amherst, NY: Prometheus Books, 2007]). Many other interesting relationships can be found involving these numbers.

OTHER COUNTRIES HAVE OTHER METHODS

Subtraction

We are not only accustomed to the arithmetic algorithms we learned in elementary school, but we often think that they are the only ways that such arithmetic can be done. This is clearly not the case. Throughout the world today, many different arithmetic algorithms—or calculating techniques—are used. We will show just a few here so that you can experience this difference.

In the United States—and some other countries—the "borrowing method" of subtraction is dominant. Until the twentieth century this method was not used in many other countries. Germany used an expansion-addition method. We can compare the subtraction methods shown in figures 2-11 and 2-12.

United States

Borrowing and Subtraction

	1	15	2	16	*Thought process*
	$\not{2}$	$\not{5}$	$\not{3}$	$\not{6}$	Borrow 10 from the 3 (3 becomes 2) and add 10 to 6 ; then $16 - 8 = \mathbf{8}$, write 8
−	1	6	2	8	$2 - 2 = \mathbf{0}$, write 0
					Borrow 10 from the 2 (2 becomes 1) and add 10 to 5; then $15 - 6 = \mathbf{9}$, write 9
		9	**0**	**8**	$1 - 1 = 0$ (0 does not need to be written)

Figure 2-11

Germany/Austria

Expansion and Addition

Thought process

```
    2   5   3   6        ? + 8 = 16, write 8, carry 1
−   1   6   2   8        ? + 1 + 2 = 3, 3 + 0 = 3; write 0
                         ? + 6 = 15, write 9, carry 1
    1           1
    ──────────────
        9   0   8        ? + 1 + 1 = 2, 2 + 0 = 2 (0 will not be written down)
```

Figure 2-12

Multiplication

In some countries—as in the United States—the factors, the multiplicand and the multiplier, are written one beneath the other (figure 2-13). In other countries they are written side by side (see figure 2-14). The partial products will be written down from right to left in some countries or vice versa in others.

United States

Factors beneath each other, multiplication denoted with "×"

```
            5   3   6        Thought process
    ×           8   7        7 × 6 = 42, write 2, carry 4
    ──────────────────        7 × 3 = 21, ...+ 4 = 25, write 5, carry 2
        3   7   5   2        7 × 5 = 35, ...+ 2 = 37, write 37
    4   2   8   8            and so on
        1       1            (then addition of the partial products)
    ──────────────────
    4   6   6   3   2
```

Figure 2-13

Germany

Factors next to each other, multiplication denoted with " · "

```
    5   3   6   ·   8   7        Thought process
        4   2   8   8   0        8·6 = 48, write 8, carry 4
            3   7   5   2        8·3 = 24, ... + 4 = 28, write 8, carry 2
                                 8·5 = 40, ... + 2 = 42, write 42
            1       1
    ──────────────────────
        4   6   6   3   2

                                 7·6 = 42, write 2, carry 4
                                 7·3 = 21, ... + 4 = 25, write 5, carry 2
                                 7·5 = 35, ... + 2 = 37, write 37
                                 (then addition of the partial products)
```

Figure 2-14

The first integer (multiplicand 536) will be multiplied with each digit of the second integer (multiplier 87) starting with the left-most digit of the multiplicand (6 of 536) and the right-most digit of the multiplier (8 of 87)—omitting zeros.[7]

For each digit of the multiplier you need a new line. You write the partial products beneath each other and add them at the end. Each partial product is written underneath the corresponding digit of the multiplier.

Asia

The multiplication method used in Asia eliminates the problem of "carrying." Each digit of the multiplicand and the multiplier is calculated individually. The product is noted in the appropriate place. If it is filled with another product, the new product will be written in a new line (see figure 2-15).

```
5  3  6  ·  8  7
   4  0  4  8          8 · 6 = 48, write 48; 8 · 3 = 24, write 24 on separate lines
      2  4  4  2       8 · 5 = 40, write 40 on the first line (in front of the 48)
            2  1       7 · 6 = 42, write 42 on the second line (as both spaces are vacant)
         3  5          7 · 3 = 21, write 21 on new line
                       7 · 5 = 35, write 35 on new line
      1     1
   ─────────────
   4  6  6  3  2       add all partial products
```

Figure 2-15

Division

Sweden

Dividend and divisor next to each other, quotient above colon as operator

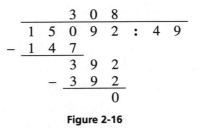

Figure 2-16

---

7. At the beginning do not omit the zero for the (indented) partial product 42,880.

The digits of the dividends and quotients will be written directly above each other—not so in Germany.

Germany

Dividend, divisor, and quotient next to each other; colon as operator

$$
\begin{array}{l}
1\ 5\ 0\ 9\ 2\ :\ 4\ 9\ =\ 3\ 0\ 8 \\
-\ 1\ 4\ 7 \\
\hline
\qquad\ \ 3\ 9\ 2 \\
\qquad -\ 3\ 9\ 2 \\
\hline
\qquad\qquad\ \ 0
\end{array}
$$

Figure 2-17

Russia

Dividend and divisor next to each other but separated by a vertical line; the quotient is written underneath a horizontal line under the divisor

$$
\begin{array}{l}
1\ 5\ 0\ 9\ 2\ \big|\ 4\ 9 \\
-\ 1\ 4\ 7\ \ \big|\ \overline{3\ 0\ 8} \\
\hline
\ \ \ 3\ 9\ 2 \\
-\ 3\ 9\ 2 \\
\hline
\ \ \ \ \ 0
\end{array}
$$

Figure 2-18

We offer these alternative techniques for arithmetic so you can appreciate that your method —one which you are very comfortable with—might be completely strange to someone raised in another country.

The Russian Peasant's Method of Multiplication

It is said that the Russian peasants used a rather strange, perhaps even primitive, method to multiply two numbers. It is actually quite simple, yet somewhat cumbersome. Let's take a look at it.

Consider the problem of finding the product of $43 \cdot 92$.

Let's work this multiplication together. We begin by setting up a chart of two columns with the two members of the product in the first row. In figure 2-19 you will see the 43 and 92 heading up the columns. One column will be formed by doubling each number to get the next, while the other column will take half the number and drop the remainder. For convenience, our first column will be the doubling column and the second column will be the halving column. Notice that by halving an odd number such as 23 (the third number in the second column) we get 11 with a remainder of 1 and we simply drop the 1. The rest of this halving process should now be clear.

43	92
86	46
172	23
344	11
688	5
1,376	2
2,752	1

Figure 2-19

Find the odd numbers in the halving column (here, the right column), then get the sum of the partner numbers in the doubling column (in this case the left column). These are highlighted in bold type. This sum gives you the originally required product of 43 and 92. In other words, with this Russian peasant's method we get $43 \cdot 92 = 172 + 344 + 688 + 2,752 = 3,956$.

In the example above, we chose to have the first column be the doubling column and the second column be the halving column. We could also have done this Russian peasant's method by halving the numbers in the first column and doubling those in the second, as in figure 2-20.

43	92
21	184
10	368
5	736
2	1,472
1	2,944

Figure 2-20

To complete the multiplication, we find the odd numbers in the halving column (in bold type), and then get the sum of their partner numbers in the second column (now the doubling column). This gives us, $43 \cdot 92 = 92 + 184 + 736 + 2,944 = 3,956$.

You are not expected to do your multiplication in this high-tech era by copying the Russian peasant's method. However, it should be fun to observe how this primitive system of arithmetic actually does work. Explorations of this kind are not only instructive but entertaining.

Here you see what was done in the above multiplication algorithm.[8]

$$
\begin{aligned}
*43 \cdot 92 &= (21 \cdot 2 + 1)(92) &= 21 \cdot 184 + & \quad \mathbf{92} = 3{,}956 \\
*21 \cdot 184 &= (10 \cdot 2 + 1)(184) &= 10 \cdot 368 + & \quad \mathbf{184} = 3{,}864 \\
10 \cdot 368 &= (5 \cdot 2 + 0)(368) &= 5 \cdot 736 + & \quad 0 = 3{,}680 \\
*5 \cdot 736 &= (2 \cdot 2 + 1)(736) &= 2 \cdot 1{,}472 + & \quad \mathbf{736} = 3{,}680 \\
2 \cdot 1{,}472 &= (1 \cdot 2 + 0)(1{,}472) &= 1 \cdot 2{,}944 + & \quad 0 = 2{,}944 \\
*1 \cdot 2{,}944 &= (0 \cdot 2 + 1)(2{,}944) &= \quad\quad 0 + & \quad \underline{\mathbf{2{,}944}} = 2{,}944
\end{aligned}
$$

Column total is 3,956

For those familiar with the binary system (i.e., base 2), one can also explain this Russian peasant's method with the following representation.

$$
\begin{aligned}
43 \cdot 92 &= (1 \cdot 2^5 + 0 \cdot 2^4 + 1 \cdot 2^3 + 0 \cdot 2^2 + 1 \cdot 2^1 + 1 \cdot 2^0) \cdot (92) \\
&= 2^0 \cdot 92 + 2^1 \cdot 92 + 2^3 \cdot 92 + 2^5 \cdot 92 \\
&= 92 + 184 + 736 + 2{,}944 \\
&= 3{,}956
\end{aligned}
$$

Whether or not you have a full understanding of the discussion of the Russian peasant's method of multiplication, you should at least now have a deeper appreciation for the multiplication algorithm you learned in school, even though most people today multiply with a calculator. There are many other multiplication algorithms. Yet the one shown here is perhaps one of the strangest, and it is through this strangeness that we can appreciate the powerful consistency of mathematics that allows us to conjure up such an algorithm.

NAPIER'S RODS

This multiplication "machine" was developed by John Napier (1550–1617), a Scottish mathematician, who was principally responsible for the development of logarithms. The device he developed consisted of flat wooden sticks with successive multiples of numbers 1–9 (see figure 2-21).

You might want to construct your own set of Napier's rods out of cardboard. Perhaps the best way to explain how to use Napier's rods is to illustrate with an example using this device.

---

8. Remember that an algorithm is a step-by-step problem-solving procedure, especially an established, recursive computational procedure for solving a problem in a finite number of steps. The asterisks indicate the bold entries in figures 2-19 and 2-20.

Figure 2-21

Consider the multiplication of 523·467. Begin by selecting the rods for 5, 2, and 3, and line them up adjacent to the index rod (see figure 2-22). Then select the appropriate rows from the index corresponding to the digits in the multiplier. Addition is done in a diagonal direction for each row.

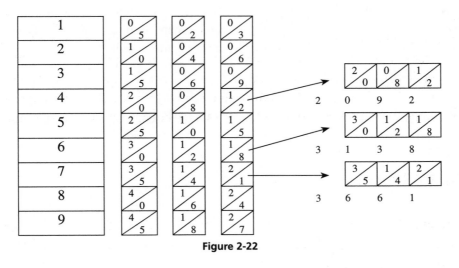

Figure 2-22

The numbers thus obtained:

$$2,092 = 4 \cdot 523$$
$$3,138 = 6 \cdot 523$$
$$3,661 = 7 \cdot 523$$

are added after considering the appropriate place values of the digits from which they were generated.

$$467 = 400 + 60 + 7$$
$$(467)(523) = (400)(523) + (60)(523) + (7)(523)$$
$$(467)(523) = 209,200$$
$$31,380$$
$$\underline{3,661}$$
$$244,241$$

A careful discussion of this last step not only will ensure a working knowledge of the computing "machine" but also should provide a thorough understanding about *why* this technique "works."

CALCULATION BY PATTERN DISCOVERY

There are times when one is faced with a daunting calculation problem that can be best solved if a pattern can be discovered. Such a pattern would nicely eliminate much tedious calculation. Here is an example of this.

Suppose you were required to find the sum of these fractions:

$$\frac{1}{2} + \frac{1}{6} + \frac{1}{12} + \frac{1}{20} + \cdots + \frac{1}{2,450}$$

To add these fractions (even with the help of a calculator) would be quite time consuming. However, if we can discover a pattern, then we can (perhaps) eliminate this tedious work.

Let's factor some of the denominators to see if a pattern evolves.

$$\frac{1}{1\cdot 2} + \frac{1}{2\cdot 3} + \frac{1}{3\cdot 4} + \frac{1}{4\cdot 5} + \cdots + \frac{1}{49\cdot 50}$$

Now let's see if the partial sums also give us a pattern

$$\frac{1}{1\cdot 2} = \frac{1}{2}$$

$$\frac{1}{1\cdot 2} + \frac{1}{2\cdot 3} = \frac{2}{3}$$

$$\frac{1}{1\cdot 2} + \frac{1}{2\cdot 3} + \frac{1}{3\cdot 4} = \frac{3}{4}$$

$$\frac{1}{1\cdot 2} + \frac{1}{2\cdot 3} + \frac{1}{3\cdot 4} + \frac{1}{4\cdot 5} = \frac{4}{5}$$

The pattern we notice is that the product in the denominator of the last fraction in the sum determines the sum. Therefore, when the series is taken to $\frac{1}{49\cdot 50}$, the sum will be $\frac{49}{50}$.

Another example where a pattern recognition is particularly useful can be seen from the following:

$$\frac{1}{\sqrt{2}+\sqrt{1}} + \frac{1}{\sqrt{3}+\sqrt{2}} + \frac{1}{\sqrt{4}+\sqrt{3}} + \cdots + \frac{1}{\sqrt{99}+\sqrt{98}} + \frac{1}{\sqrt{100}+\sqrt{99}}$$

To begin, we would look for a pattern. To determine if a pattern can be found, we shall convert each of the fractions to their equivalent with a rational denominator. We do this by multiplying each fraction by 1, in the form of the denominator's conjugate:[9]

$$\frac{1}{\sqrt{2}+\sqrt{1}} = \frac{1}{\sqrt{2}+\sqrt{1}} \cdot \frac{\sqrt{2}-\sqrt{1}}{\sqrt{2}-\sqrt{1}} = \frac{\sqrt{2}-\sqrt{1}}{2-1}$$

$$\frac{1}{\sqrt{3}+\sqrt{2}} = \frac{1}{\sqrt{3}+\sqrt{2}} \cdot \frac{\sqrt{3}-\sqrt{2}}{\sqrt{3}-\sqrt{2}} = \frac{\sqrt{3}-\sqrt{2}}{3-2}$$

$$\frac{1}{\sqrt{4}+\sqrt{3}} = \frac{1}{\sqrt{4}+\sqrt{3}} \cdot \frac{\sqrt{4}-\sqrt{3}}{\sqrt{4}-\sqrt{3}} = \frac{\sqrt{4}-\sqrt{3}}{4-3}$$

Continuing to:

$$\frac{1}{\sqrt{100}+\sqrt{99}} = \frac{1}{\sqrt{100}+\sqrt{99}} \cdot \frac{\sqrt{100}-\sqrt{99}}{\sqrt{100}-\sqrt{99}} = \frac{\sqrt{100}-\sqrt{99}}{100-99}$$

The series can then be written as:

$$\frac{1}{\sqrt{2}+\sqrt{1}} + \frac{1}{\sqrt{3}+\sqrt{2}} + \frac{1}{\sqrt{4}+\sqrt{3}} + \ldots + \frac{1}{\sqrt{99}+\sqrt{98}} + \frac{1}{\sqrt{100}+\sqrt{99}}$$

$$= \frac{\sqrt{2}-\sqrt{1}}{2-1} + \frac{\sqrt{3}-\sqrt{2}}{3-2} + \frac{\sqrt{4}-\sqrt{3}}{4-3} + \ldots + \frac{\sqrt{99}-\sqrt{98}}{99-98} + \frac{\sqrt{100}-\sqrt{99}}{100-99}$$

$$= \sqrt{2}-\sqrt{1} + \sqrt{3}-\sqrt{2} + \sqrt{4}-\sqrt{3} + \ldots + \sqrt{99}-\sqrt{98} + \sqrt{100}-\sqrt{99}$$

$$= -\sqrt{1} + \sqrt{100} = 10-1 = 9$$

So, before you embark on any tedious calculations, be sure to first see if you can find a pattern or transform the given information into a useful pattern.

ALPHAMETICS

One of the great strides made by Western civilization (which was adopted from the Indian and then Arabic civilizations) was the use of a place-value system for our arithmetics. Working with Roman numerals was not only cumbersome but made many algorithms nearly impossible. The first appearance of the Hindu-Arabic numerals was in Fibonacci's book *Liber abaci* in 1202. Beyond its usefulness, the place-value system can also provide

---

9. The *conjugate* of a number in the form $\sqrt{a}-\sqrt{b}$ is $\sqrt{a}+\sqrt{b}$, and vice versa.

us with some recreational mathematics that can stretch our understanding and facility with the algorithms we use with it.

Applying reasoning skills to analyzing an addition algorithm situation can be very useful in sharpening your mathematical thinking. *Alphametics* (or *verbal arithmetics*, also known as *cryptarithmetics*, *crypt-arithmetics*, or *cryptarithms*) are mathematical puzzles that appear in several disguises. Sometimes the problems are associated with the restoration of digits in a computational problem; at other times, the problem is associated with decoding the complete arithmetical problem, where letters of the alphabet represent all the digits. Basically, construction of this type of puzzle is not difficult, but the solution requires a thorough investigation of many arithmetic elements. Every clue must be tested in all phases of the problem and carefully followed up.

There are two fairly obvious (but worth stating) rules that every alphametic obeys:

1. The mapping of letters to numbers is one to one. That is, the same letter always stands for the same digit, and the same digit is always represented by the same letter.
2. The digit zero is not allowed to appear as the left-most digit in any of the addends or the sum.

For example, suppose we were to eliminate certain digits in an addition problem. We could be asked to supply the missing digits. Let us also assume that we do not know what these digits are. We may then be left with the following skeleton problem (figure 2-23):

	①	②	③	④	⑤
			__	6	2
		3	9	4	__
+		__	8	__	7
__		3	3	1	2

Figure 2-23

From column five, $2 + \underline{\hspace{0.5cm}} + 7 = 12$ (this will give us the digit 2 that we seek). Therefore, the missing digit in the fifth column must be 3. In the fourth column, we have $1 + 6 + 4 + \underline{\hspace{0.5cm}} = 11$ (to get the units digit 1), therefore the missing digit must be zero.

In the third column, we have $1 + \underline{\hspace{0.5cm}} + 9 + 8 = 23$, and the missing digit must be 5.

Now, from the second column, we have now $2 + 3 + \underline{\hspace{0.5cm}} = 13$. This implies that the digit must be 8, and therefore, the digit to the left of 3 in the first column, bottom row must be 1.

Thus, we have reconstructed the problem (figure 2-24).

```
   ①    ②    ③    ④    ⑤
              5    6    2
         3    9    4    3
    +    8    8    0    7
   ─────────────────────
    1    3    3    1    2
```

Figure 2-24

You should now be able to find the missing digits in the addition problem shown in figure 2- 25.

(Don't look to the end of this section just yet.)

```
         5    6    7   __
    +   __    8   __    9
   ─────────────────────
   __    3   __    3    3
```

Figure 2-25

The following example (figure 2-26) will show a problem that has more than one solution.

```
        __    8    7
         3   __    1
    +    5    6   __
   ──────────────────
    __    3   __    0
```

Figure 2-26

In the units column, $7 + 1 + \_\_ = 10$, so the missing digit must be **2**.

```
        __    8    7
         3   __    1
    +    5    6    2
   ──────────────────
    __    3   __    0
```

Figure 2-27

In the tens column (figure 2-27), $1 + 8 + \_\_ + 6 = \_\_$, or $15 + \_\_ = \_\_$. An inspection must now be made of the hundreds column so that all possible outcomes are considered. In the hundreds column, we have $\_\_ + 3 + 5 = 13$. Thus, if we assigned any of the digits 5, 6, 7, 8, or 9 for the value of the missing number (second row) in the tens column, we would have $15 + 5 = 20$, or $15 + 6 = 21$, or $15 + 7 = 22$, or $15 + 8 = 23$, or $15 + 9 = 24$.

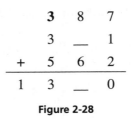

Figure 2-28

This will then make the digit in the hundreds column equal to **3**, since a 2 is being carried. Hence, we have as possible solutions:

387		387		387		387		387
351		361		371		381		391
+ 562	or	+ 562	or	+ 562	or	+ 562	or	+ 562
1,300		1,310		1,320		1,330		1,340

On the other hand, if we were to assign values for the missing digit in the second row of the tens column to be 0, 1, 2, 3, or 4, then the digit in the first row of the hundreds column would have to be a 4, since 1 is now being carried over from the tens column (rather than the 2 as before). These additional solutions would be acceptable:

487		487		487		487		487
301		311		321		331		341
+ 562	or	+ 562	or	+ 562	or	+ 562	or	+ 562
1,350		1,360		1,370		1,380		1,390

Therefore, there are ten different solutions having two missing digits in the same column.

In a second type of problem, where all digits are represented by letters (hence the name *alphametics*), the problem is quite different from the preceding ones. Here, the clues from the "puzzle" must be analyzed for all different possible values to be assigned to the letters. No general rule can be given for the solution of alphametic problems. What is required is an understanding of basic arithmetic, logical reasoning, and plenty of patience.

One fine example of this type is the addition problem shown in figure 2-29.

①	②	③	④	⑤
F	O	R	T	Y
		T	E	N
+		T	E	N
S	I	X	T	Y

Figure 2-29

Since the first row and the fourth row retain the **T** and **Y** in the sum, this would imply that the sum of both the **E**s and the **N**s in columns four and five must end in zero. If we let **N** = 0, then **E** must equal 5, and 1 is carried over to column three. We now have the following figure 2-30):

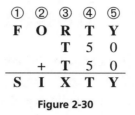

Figure 2-30

Since there are two spaces before each **T E N**, the **O** in **F O R T Y** must be 9, and with 2 carried over from the hundreds place (column three), the **I** must be 1. And a 1 is carried to column one, making **F** + 1 = **S**. Do you know why 2 and not 1 was carried over to the second column? The reason 2 must be carried from column three is that if a 1 were carried, the digits **I** and **N** would both be zero. We are now left with the numbers 2, 3, 4, 6, 7, and 8 unassigned.

①	②	③	④	⑤
F	9	R	T	Y
			T	5	0
+		T	5	0
S	1	X	T	Y

Figure 2-31

In the hundreds column (figure 2-31), we have 2**T** + **R** + 1 (the 1 being carried over from column four), whose sum must be equal to, or greater than, 22, which implies **T** and **R** must be greater than 5. Therefore, **F** and **S** will each be one of 2, 3, or 4. Now **X** cannot be equal to 3; otherwise, **F** and **S** would not be consecutive numbers. Then **X** equals 2 or 4, which is impossible if **T** is equal to or less than 7. Hence **T** must be 8, with **R** equal to 7 and **X** equal to 4. Then **F** = 2 and **S** = 3, leaving **Y** = 6. Hence the solution to the problem is

```
   2   9   7   8   6
               8   5   0
   +       8   5   0
   3   1   4   8   6
```

By now you should have your technique well in hand. As before, the following letters represent the digits of a simple addition. Here is one of the more popular—yet challenging—alphametic problems.

A classic example, published in the July 1924 issue of *Strand Magazine*[10] by Henry Dudeney (1857–1930), is:

$$\begin{array}{r} \text{S E N D} \\ +\text{M O R E} \\ \hline \text{M O N E Y} \end{array}$$

Find the digits that represent the letters to make this addition correct. You might also want to show that the solution is unique.

Most important in this activity is the analysis, and you should pay particular attention to the reasoning used. At first, this may appear to be a daunting task, but if you take it step by step, you will find it rewarding and entertaining.

The sum of two four-digit numbers cannot yield a number greater than 19,999. Therefore **M = 1**.

We then have MORE < 2,000 and SEND < 10,000. It follows that MONEY < 11,999. Thus, **O** can be either 0 or 1. But the 1 is already used; therefore, **O = 0**.

We now have:

$$\begin{array}{r} \text{S E N D} \\ +\text{1 0 R E} \\ \hline \text{1 0 N E Y} \end{array}$$

Now MORE < 1,100. If SEND were less than 9,000, then MONEY < 10,100, which would imply that N = 0. But this cannot be since 0 was already used; therefore SEND > 9,000, so that **S = 9**.

We now have:

$$\begin{array}{r} \text{9 E N D} \\ +\text{1 0 R E} \\ \hline \text{1 0 N E Y} \end{array}$$

The remaining digits from which we may complete the problem are 2, 3, 4, 5, 6, 7, and 8.

Let us examine the units digits. The greatest sum (of the remaining digits) is 7 + 8 = 15 and the least sum is 2 + 3 = 5.

---

10. *Strand Magazine* 68 (July 1924): 97 and 214.

If D + E < 10, then D + E = Y, with no carryover into the tens column. Otherwise D + E = Y + 10, with a 1 carried over to the tens column. Since there are no obvious answers at this point, we will assume no carryover and see where this takes us.

Using this argument and moving one step further to the tens column, we get N + R = E, with no carryover, or N + R = E + 10, with a carryover of 1 to the hundreds column. However, if there is no carryover to the hundreds column, then E + 0 = N, which implies that E = N. This is not permissible. Therefore, there must be a carryover to the hundreds column. So N + R = E + 10, and E + 0 + 1 = N, or E + 1 = N.

Substituting this value for N into the previous equation we get: (E + 1) + R = E + 10, which implies that R = 9. But this has already been used for the value of S. We must try a different approach.

We shall assume, therefore, that D + E = Y + 10, since we apparently need a carryover into the tens column, where we just reached a dead end.

Now the sum in the tens column is 1 + 2 + 3 < 1 + N + R < 1 + 7 + 8. If, however, 1 + N + R < 10, there will be no carryover to the hundreds column, leaving the previous dilemma of E = N, which is not allowed. We then have 1 + N + R = E + 10, which ensures the needed carryover to the hundreds column.

Therefore, 1 + E + 0 = N, or E + 1 = N.

Substituting this in the above equation (1 + N + R = E + 10) gives us 1 + (E + 1) + R = E + 10, or **R = 8**.

We now have:

$$9\,E\,N\,D$$
$$+\,1\,0\,8\,E$$
$$1\,0\,N\,E\,Y$$

From the remaining list of available digits, we find that D + E < 14.

So from the equation D + E = Y + 10, Y is either 2 or 3. If Y = 3, then D + E = 13, implying that the digits D and E can take on only 6 or 7.

If D = 6 and E = 7, then from the previous equation E + 1 = N, we would have N = 8, which is unacceptable since R = 8.

If D = 7 and E = 6, then from the previous equation E + 1 = N, we would have N = 7, which is unacceptable since D = 7. Therefore, **Y = 2**.

We now have:

$$9\,E\,N\,D$$
$$+\,1\,0\,8\,E$$
$$1\,0\,N\,E\,2$$

Thus, D + E = 12. The only way to get this sum is with 5 and 7. We have already determined that E cannot be 7; therefore, **D = 7** and **E = 5**. We can now again use the equation E + 1 = N to get **N = 6**.

Finally we get the solution:

$$9\,5\,6\,7$$
$$+1\,0\,8\,5$$
$$10\,6\,5\,2$$

This rather strenuous activity should provide some refreshing exercise in strengthening one's facility in mathematics.

Here are two more alphametics problems with their respective solutions:

APPLE + ORANGE = BANANA

There are two solutions, since L and G are interchangeable.

85,524 + 698,314 = 783,838
85,514 + 698,324 = 783,838

The same as

FERMAT · S = LAST + THEOREM
703,612 · 4 = 5,142 + 2,809,306
703,612 · 4 = 9,142 + 2,805,306

You may wish to try some more alphametic problems. Here is one that has ten possible solutions:

ALLS + WELL + THAT + ENDS +WELL = SWELL

1002+4700+6316+7982+4700=24700, where A=1, D=8, E=7, H=3, L=0, N=9, S=2, T=6, W=4
1002+4700+6916+7382+4700=24700, where A=1, D=8, E=7, H=9, L=0, N=3, S=2 T=6, W=4
4002+1700+6346+7952+1700=21700, where A=4, D=5, E=7, H=3, L=0, N=9, S=2, T=6, W=1
4002+1700+6946+7352+1700=21700, where A=4, D=5, E=7, H=9, L=0, N=3, S=2, T=6, W=1
7003+9800+4574+8623+9800=39800, where A=7, D=2, E=8, H=5, L=0, N=6, S=3, T=4, W=9
7003+9800+4674+8523+9800=39800, where A=7, D=2, E=8, H=6, L=0, N=5, S=3, T=4, W=9
6552+7455+1061+4932+7455=27455, where A=6, D=3, E=4, H=0, L=5, N=9, S=2, T=1, W=7
6552+7455+1961+4032+7455=27455, where A=6, D=3, E=4, H=9, L=5, N=0, S=2, T=1, W=7
4663+9766+8048+7523+9766=39766, where A=4, D=2, E=7, H=0, L=6, N=5, S=3, T=8, W=9
4663+9766+8548+7023+9766=39766, where A=4, D=2, E=7, H=5, L=6, N=0, S=3, T=8, W=9

Despite the ubiquity of the calculator, there is much to be gained from a deeper under-standing of arithmetic relationships. This brief ride through some of the unusual (and surpris-ing) aspects of arithmetic should not only allow you to appreciate the algorithms you have learned to master in elementary school, but should also give you a finer insight—not to men-tion some newfound skills—into this delightful and engrossing aspect of mathematics.

The completed solution to the addition problem on page 89 is:

$$
\begin{array}{r r r r r}
 & 5 & 6 & 7 & \_ \\
+\ & \_ & 8 & \_ & 9 \\
\hline
\_\ & 3 & \_ & 3 & 3 \\
\end{array}
$$

$$
\begin{array}{r r r r r}
 & 5 & 6 & 7 & ④ \\
+\ & ⑦ & 8 & ⑤ & 9 \\
\hline
① & 3 & ⑤ & 3 & 3 \\
\end{array}
$$

Chapter 3

ARITHMETIC LOOPS

The frustration of struggling to solve a problem—both in mathematics and in everyday-life situations—may make you feel as though you are "going around in circles." There are, interestingly enough, phenomena in mathematics where "going around in circles" is searched for and, when discovered, very much appreciated. In this chapter, we will explore a wide variety of truly astounding aspects of our number system. These surprising number relationships are ones that will form loops—analogous to "going around in circles." By that we mean that we begin with a number and then perform some consistent procedure, which, after several iterations, will bring us back to the starting point. That is, these procedures operated on these numbers will eventually lead us into a surprising loop that will not take us any further. It just repeats itself or results in a dead end! We shall begin our journey through this aspect of mathematics with one example of a loop involving the number 89.

THE 89 LOOP

Let's see what happens when you take any number and get the sum of its digits, and continue to take each successive sum and further sum the digits until you end up with a single-digit number. It is clearly no surprise that you will eventually end up with any one of the natural numbers: 1, 2, 3, 4, 5, 6, 7, 8, or 9.

For example, we will consider two cases:

$$n = 985: \quad 9 + 8 + 5 = 22, 2 + 2 = 4$$
$$n = 127: \quad 1 + 2 + 7 = 10, 1 + 0 = 1$$

Yet, if we modify this process just a bit, we get a very different result. Let's see what happens when you take any number and get the sum of *the squares* of its digits, then continue this process of finding the sum of the squares of the digits of this resulting number and so on. Each time, curiously enough, you will eventually reach 1 or 89. Take a look at some examples that follow.

We will begin with a randomly selected number, say, 5. We will find the sum of the squares of the digits of the number 5, squared, and continue the process, each time finding the sum of the squares of the digits of the resulting number.

Example 1: $n = 5$

$5^2 = 25, 2^2 + 5^2 = 29, 2^2 + 9^2 = 85, 8^2 + 5^2 = \mathbf{89}, 8^2 + 9^2 = 145, 1^2 + 4^2 + 5^2 = 42,$

$4^2 + 2^2 = 20, 2^2 + 0^2 = 4, 4^2 = 16, 1^2 + 6^2 = 37, 3^2 + 7^2 = 58, 5^2 + 8^2 = \mathbf{89}, \ldots$

Once we have reached 89, you will notice that we got into what we call a *loop*, since we always seem to get back to the number 89 with continuous repetitions of this process. We landed in a loop that looks like this:

[89, 145, 42, 20, 4, 16, 37, 58, (89)] (see figure 3-1).

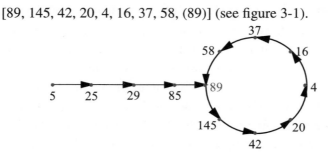

Figure 3-1

Let's try this procedure with another starting number, say, 30.

Example 2: $n = 30$

$3^2 + 0^2 = 9, 9^2 = 81, 8^2 + 1^2 = 65, 6^2 + 5^2 = 61, 6^2 + 1^2 = 37, 3^2 + 7^2 = 58,$

$5^2 + 8^2 = \mathbf{89}, 8^2 + 9^2 = 145, 1^2 + 4^2 + 5^2 = 42, 4^2 + 2^2 = 20, 2^2 + 0^2 = 4,$

$4^2 = 16, 1^2 + 6^2 = 37, 3^2 + 7^2 = 58, 5^2 + 8^2 = \mathbf{89}, \ldots$

We actually entered the loop when we arrive at 37.

[30, 9, 81, 65, 61, ***37, 58, 89, 145, 42, 20, 4, 16, 37***, **58** (, **89**)][1] (see figure 3-2).

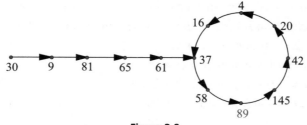

Figure 3-2

---

1. Bold entries are members of a loop. Bold italic indicates the beginning of a loop.

Let's try this procedure with the number 13.

Example 3: $n = 13$

$$1^2 + 3^2 = 10, \ 1^2 + 0^2 = \mathbf{1}, \ 1^2 + 0^2 = \mathbf{1}, \ \ldots$$

When we reach the number 1 a loop is formed, getting us back to 1, over and over. Here we land in a loop that we can denote as [1 (, 1)] of length 1 (see figure 3-3).

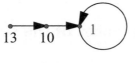

Figure 3-3

We shall now try 32,

$$3^2 + 2^2 = 13, \ 1^2 + 3^2 = 10, \ 1^2 + 0^2 = \mathbf{1}, \ 1^2 = \mathbf{1}, \ \ldots$$

Again, notice that when we reach 1, we can't go further, since we will always get 1. Now let's try this procedure on the number 33:

$$3^2 + 3^2 = 18, \ 1^2 + 8^2 = 65, \ 6^2 + 5^2 = 61, \ 6^2 + 1^2 = 37, \ 3^2 + 7^2 = 58, \ 5^2 + 8^2 = \mathbf{89},$$
$$8^2 + 9^2 = 145, \ 1^2 + 4^2 + 5^2 = 42, \ 4^2 + 2^2 = 20, \ 2^2 + 0^2 = 4, \ 4^2 = 16, \ 1^2 + 6^2 = 37,$$
$$3^2 + 7^2 = 58, \ 5^2 + 8^2 = \mathbf{89}, \ \ldots$$

Remember, when we reach 89, we begin to enter into the loop.

Let's take a larger number, say, 80, and try to apply the process on this number:

$$8^2 + 0^2 = 64, \ 6^2 + 4^2 = 52, \ 5^2 + 2^2 = 29, \ 2^2 + 9^2 = 85, \ 8^2 + 5^2 = \mathbf{89}, \ 8^2 + 9^2 = 145,$$
$$1^2 + 4^2 + 5^2 = 42, \ 4^2 + 2^2 = 20, \ 2^2 + 0^2 = 4, \ 4^2 = 16, \ 1^2 + 6^2 = 37, \ 3^2 + 7^2 = 58,$$
$$5^2 + 8^2 = \mathbf{89}, \ \ldots$$

Remember, when we reach 89 or 1, you have essentially finished, since you will then enter a loop.

Suppose we try this process on the next larger integer, 81:

$$8^2 + 1^2 = 65, \ 6^2 + 5^2 = 61, \ 6^2 + 1^2 = 37, \ 3^2 + 7^2 = 58, \ 5^2 + 8^2 = \mathbf{89}, \ 8^2 + 9^2 = 145,$$
$$1^2 + 4^2 + 5^2 = 42, \ 4^2 + 2^2 = 20, \ 2^2 + 0^2 = 4, \ 4^2 = 16, \ 1^2 + 6^2 = 37, \ 3^2 + 7^2 = 58,$$
$$5^2 + 8^2 = \mathbf{89}, \ \ldots$$

Now when we try it for 82, we quickly get into the loop.

$$8^2 + 2^2 = 68, \ 6^2 + 8^2 = 100, \ 1^2 + 0^2 + 0^2 = \mathbf{1}, \ 1^2 = \mathbf{1}, \ \ldots$$

And when we try it for 85, we get 89 immediately:

$$8^2 + 5^2 = \mathbf{89}, \ 8^2 + 9^2 = 145, \ 1^2 + 4^2 + 5^2 = 42, \ 4^2 + 2^2 = 20, \ 2^2 + 0^2 = 4, \ 4^2 = 16,$$
$$1^2 + 6^2 = 37, \ 3^2 + 7^2 = 58, \ 5^2 + 8^2 = \mathbf{89}, \ \ldots$$

Notice that for each of the numbers we selected, you end up with either 1 or 89 and then, of course, you get into a loop that brings you right back to 1 or 89.

By now you should realize that certain numbers follow a path to the loop. The following figures will diagram these various routes for the 89 loop and the 1 loop.

Visualization of the sequences of the sums of the squares of digits

For $n = 2, 3, 4, 5, 6, 8, 9, 11, 12, 14, \ldots, 18, 20, 21, 22, 24, \ldots, 27, 29, 30, 33, \ldots, 43, 45, \ldots$ with the loop [89, 145, 42, 20, 4, 16, 37, 58].

See figure 3-4 and the following table.

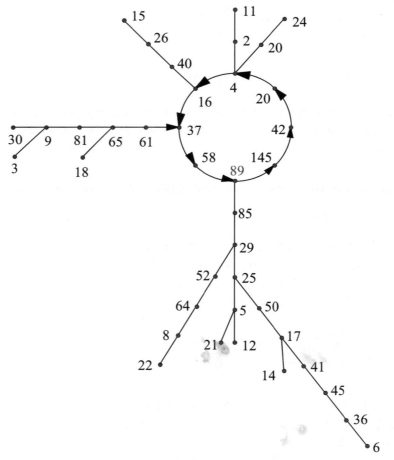

Figure 3-4

For n = 1, 7, 10, 13, 19, 23, 28, 31, 32, 44, 49, 68, 70, 79, 82, 86, 91, 94, 97, 100 with the loop [1] = [1 (, 1)] (see figure 3-5).

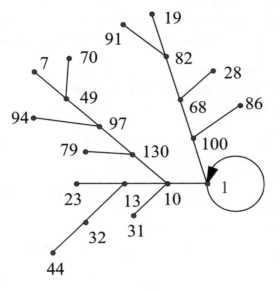

Figure 3-5

The following is a summary of the paths all of the numbers from 1 to 100 take to reach their respective loops. As you can see, the size of the number is not related to the length of the path to the loop.

Sequences of the sums of the squares of the digits for n = 1, 2, ... , 100

n	Sequence of the sums of the squares of digits
1	[**1** (, **1**)]
2	[2, *4, 16, 37, 58*, **89, 145, 42, 20, 4, 16, 37, 58** (, **89**)]
3	[3, 9, 81, 65, 61, *37, 58*, **89, 145, 42, 20, 4, 16,** *37, 58* (, **89**)]
4	[*4, 16, 37, 58*, **89, 145, 42, 20, 4, 16, 37, 58** (, **89**)]
5	[5, 25, 29, 85, **89, 145, 42, 20, 4, 16, 37, 58** (, **89**)]
6	[6, 36, 45, 41, 17, 50, 25, 29, 85, **89, 145, 42, 20, 4, 16, 37, 58** (, **89**)]
7	[7, 49, 97, 130, 10, **1** (, **1**)]
8	[8, 64, 52, 29, 85, **89, 145, 42, 20, 4, 16, 37, 58** (, **89**)]
9	[9, 81, 65, 61, *37, 58*, **89, 145, 42, 20, 4, 16,** *37, 58* (, **89**)]
10	[10, **1** (, **1**)]
11	[11, 2, *4, 16, 37, 58*, **89, 145, 42, 20, 4, 16, 37, 58** (, **89**)]
12	[12, 5, 25, 29, 85, **89, 145, 42, 20, 4, 16, 37, 58** (, **89**)]
13	[13, 10, **1** (, **1**)]
14	[14, 17, 50, 25, 29, 85, **89, 145, 42, 20, 4, 16, 37, 58** (, **89**)]

n	Sequence of the sums of the squares of digits
15	[15, 26, 40, *16*, *37*, *58*, **89**, **145**, **42**, **20**, **4**, *16*, **37**, **58** (, **89**)]
16	[*16*, *37*, *58*, **89**, **145**, **42**, **20**, **4**, *16*, **37**, **58** (, **89**)]
17	[17, 50, 25, 29, 85, **89**, **145**, **42**, **20**, **4**, **16**, **37**, **58** (, **89**)]
18	[18, 65, 61, *37*, *58*, **89**, **145**, **42**, **20**, **4**, **16**, *37*, **58** (, **89**)]
19	[19, 82, 68, 100, **1** (, **1**)]
20	[*20*, *4*, *16*, *37*, *58*, **89**, **145**, **42**, *20*, **4**, **16**, **37**, **58** (, **89**)]
21	[21, 5, 25, 29, 85, **89**, **145**, **42**, **20**, **4**, **16**, **37**, **58** (, **89**)]
22	[22, 8, 64, 52, 29, 85, **89**, **145**, **42**, **20**, **4**, **16**, **37**, **58** (, **89**)]
23	[23, 13, 10, **1** (, **1**)]
24	[24, *20,* *4*, *16*, *37*, *58*, **89**, **145**, **42**, *20*, **4**, **16**, **37**, **58** (, **89**)]
25	[25, 29, 85, **89**, **145**, **42**, **20**, **4**, **16**, **37**, **58** (, **89**)]
26	[26, 40, 16, *37,* *58*, **89**, **145**, **42**, **20**, **4**, **16**, *37*, **58** (, **89**)]
27	[27, 53, 34, 25, 29, 85, **89**, **145**, **42**, **20**, **4**, **16**, **37**, **58** (, **89**)]
28	[28, 68, 100, **1** (, **1**)]
29	[29, 85, **89**, **145**, **42**, **20**, **4**, **16**, **37**, **58** (, **89**)]
30	[30, 9, 81, 65, 61, *37*, *58*, **89**, **145**, **42**, **20**, **4**, **16**, *37*, **58** (, **89**)]
31	[31, 10, **1** (, **1**)]
32	[32, 13, 10, **1** (, **1**)]
33	[33, 18, 65, 61, *37*, *58*, **89**, **145**, **42**, **20**, **4**, **16**, *37*, **58** (, **89**)]
34	[34, 25, 29, 85, **89**, **145**, **42**, **20**, **4**, **16**, **37**, **58** (, **89**)]
35	[35, 34, 25, 29, 85, **89**, **145**, **42**, **20**, **4**, **16**, **37**, **58** (, **89**)]
36	[36, 45, 41, 17, 50, 25, 29, 85, **89**, **145**, **42**, **20**, **4**, **16**, **37**, **58** (, **89**)]
37	[*37*, *58*, **89**, **145**, **42**, **20**, **4**, **16**, *37*, **58** (, **89**)]
38	[38, 73, *58*, **89**, **145**, **42**, **20**, **4**, **16**, **37**, *58* (, **89**)]
39	[39, 90, 81, 65, 61, *37*, *58*, **89**, **145**, **42**, **20**, **4**, **16**, *37*, **58** (, **89**)]
40	[40, *16*, *37*, *58*, **89**, **145**, **42**, **20**, **4**, *16*, **37**, **58** (, **89**)]
41	[41, 17, 50, 25, 29, 85, **89**, **145**, **42**, **20**, **4**, **16**, **37**, **58** (, **89**)]
42	[*42*, *20*, *4*, *16*, *37*, *58*, **89**, **145**, *42*, **20**, **4**, **16**, **37**, **58** (, **89**)]
43	[43, 25, 29, 85, **89**, **145**, **42**, **20**, **4**, **16**, **37**, **58** (, **89**)]
44	[44, 32, 13, 10, **1** (, **1**)]
45	[45, 41, 17, 50, 25, 29, 85, **89**, **145**, **42**, **20**, **4**, **16**, **37**, **58** (, **89**)]
46	[46, 52, 29, 85, **89**, **145**, **42**, **20**, **4**, **16**, **37**, **58** (, **89**)]
47	[47, 65, 61, *37*, *58*, **89**, **145**, **42**, **20**, **4**, **16**, *37*, **58** (, **89**)]
48	[48, 80, 64, 52, 29, 85, **89**, **145**, **42**, **20**, **4**, **16**, **37**, **58** (, **89**)]
49	[49, 97, 130, 10, **1** (, **1**)]
50	[50, 25, 29, 85, **89**, **145**, **42**, **20**, **4**, **16**, **37**, **58** (, **89**)]
51	[51, 26, 40, *16*, *37*, *58*, **89**, **145**, **42**, **20**, **4**, **16**, **37**, **58** (, **89**)]
52	[52, 29, 85, **89**, **145**, **42**, **20**, **4**, **16**, **37**, **58** (, **89**)]
53	[53, 34, 25, 29, 85, **89**, **145**, **42**, **20**, **4**, **16**, **37**, **58** (, **89**)]
54	[54, 41, 17, 50, 25, 29, 85, **89**, **145**, **42**, **20**, **4**, **16**, **37**, **58** (, **89**)]
55	[55, 50, 25, 29, 85, **89**, **145**, **42**, **20**, **4**, **16**, **37**, **58** (, **89**)]

n	Sequence of the sums of the squares of digits
56	[56, 61, *37*, *58*, **89**, **145**, **42**, **20**, **4**, **16**, *37*, **58** (, **89**)]
57	[57, 74, 65, 61, *37*, *58*, **89**, **145**, **42**, **20**, **4**, **16**, *37*, **58** (, **89**)]
58	[*58*, **89**, **145**, **42**, **20**, **4**, **16**, **37**, *58* (, **89**)]
59	[59, 106, *37*, *58*, **89**, **145**, **42**, **20**, **4**, **16**, *37*, **58** (, **89**)]
60	[60, 36, 45, 41, 17, 50, 25, 29, 85, **89**, **145**, **42**, **20**, **4**, **16**, **37**, **58** (, **89**)]
61	[61, *37*, *58*, **89**, **145**, **42**, **20**, **4**, **16**, *37*, **58** (, **89**)]
62	[62, 40, *16*, *37*, *58*, **89**, **145**, **42**, **20**, **4**, *16*, **37**, **58** (, **89**)]
63	[63, 45, 41, 17, 50, 25, 29, 85, **89**, **145**, **42**, **20**, **4**, **16**, **37**, **58** (, **89**)]
64	[64, 52, 29, 85, **89**, **145**, **42**, **20**, **4**, **16**, **37**, **58** (, **89**)]
65	[65, 61, *37*, *58*, **89**, **145**, **42**, **20**, **4**, **16**, *37*, **58** (, **89**)]
66	[66, 72, 53, 34, 25, 29, 85, **89**, **145**, **42**, **20**, **4**, **16**, **37**, **58** (, **89**)]
67	[67, 85, **89**, **145**, **42**, **20**, **4**, **16**, **37**, **58** (, **89**)]
68	[68, 100, **1** (, **1**)]
69	[69, 117, 51, 26, 40, *16*, *37*, *58*, **89**, **145**, **42**, **20**, **4**, *16*, **37**, **58** (, **89**)]
70	[70, 49, 97, 130, 10, **1** (, **1**)]
71	[71, 50, 25, 29, 85, **89**, **145**, **42**, **20**, **4**, **16**, **37**, **58** (, **89**)]
72	[72, 53, 34, 25, 29, 85, **89**, **145**, **42**, **20**, **4**, **16**, **37**, **58** (, **89**)]
73	[73, *58*, **89**, **145**, **42**, **20**, **4**, **16**, **37**, *58* (, **89**)]
74	[74, 65, 61, *37*, *58*, **89**, **145**, **42**, **20**, **4**, **16**, *37*, **58** (, **89**)]
75	[75, 74, 65, 61, *37*, *58*, **89**, **145**, **42**, **20**, **4**, **16**, *37*, **58** (, **89**)]
76	[76, 85, **89**, **145**, **42**, **20**, **4**, **16**, **37**, **58** (, **89**)]
77	[77, 98, *145*, **42**, 20, *4*, *16*, *37*, *58*, **89**, *145*, **42**, **20**, **4**, **16**, **37**, **58** (, **89**)]
78	[78, 113, 11, 2, *4*, *16*, *37*, *58*, **89**, **145**, **42**, **20**, *4*, **16**, **37**, **58** (, **89**)]
79	[79, 130, 10, **1** (, **1**)]
80	[80, 64, 52, 29, 85, **89**, **145**, **42**, **20**, **4**, **16**, **37**, **58** (, **89**)]
81	[81, 65, 61, *37*, *58*, **89**, **145**, **42**, **20**, **4**, **16**, *37*, **58** (, **89**)]
82	[82, 68, 100, **1** (, **1**)]
83	[83, 73, *58*, **89**, **145**, **42**, **20**, **4**, **16**, **37**, *58* (, **89**)]
84	[84, 80, 64, 52, 29, 85, **89**, **145**, **42**, **20**, **4**, **16**, **37**, **58** (, **89**)]
85	[85, **89**, **145**, **42**, **20**, **4**, **16**, **37**, **58** (, **89**)]
86	[86, 100, **1** (, **1**)]
87	[87, 113, 11, 2, *4*, *16*, *37*, *58*, **89**, **145**, **42**, **20**, *4*, **16**, **37**, **58** (, **89**)]
88	[88, 128, 69, 117, 51, 26, 40, *16*, *37*, *58*, **89**, **145**, **42**, **20**, **4**, *16*, **37**, **58** (, **89**)]
89	[**89**, **145**, **42**, **20**, **4**, **16**, **37**, **58** (, **89**)]
90	[90, 81, 65, 61, *37*, *58*, **89**, **145**, **42**, **20**, **4**, **16**, *37*, **58** (, **89**)]
91	[91, 82, 68, 100, **1** (, **1**)]
92	[92, 85, **89**, **145**, **42**, **20**, **4**, **16**, **37**, **58** (, **89**)]
93	[93, 90, 81, 65, 61, *37*, *58*, **89**, **145**, **42**, **20**, **4**, **16**, *37*, **58** (, **89**)]
94	[94, 97, 130, 10, **1** (, **1**)]
95	[95, 106, *37*, *58*, **89**, **145**, **42**, **20**, **4**, **16**, *37*, **58** (, **89**)]
96	[96, 117, 51, 26, 40, *16*, *37*, *58*, **89**, **145**, **42**, **20**, **4**, *16*, **37**, **58** (, **89**)]

n	Sequence of the sums of the squares of digits
97	[97, 130, 10, **1** (, **1**)]
98	[98, *145*, *42*, *20*, *4*, *16*, *37*, *58*, **89**, *145*, *42*, *20*, *4*, *16*, *37*, *58* (, **89**)]
99	[99, 162, 41, 17, 50, 25, 29, 85, **89**, **145**, **42**, **20**, **4**, **16**, **37**, **58** (, **89**)]
100	[100, **1** (, **1**)]

We can notice that for all $n \le 100$, either the loop [89, 145, 42, 20, 4, 16, 37, 58] or the loop [1] will evolve. Yet, also for $9{,}990 \le n \le 10{,}000$, one of these two loops will materialize, as you can see from the table below.

n	Sequence of the sums of the squares of digits
9,990	[9,990, 243, 29, 85, **89**, **145**, **42**, **20**, **4**, **16**, **37**, **58** (, **89**)]
9,991	[9,991, 244, 36, 45, 41, 17, 50, 25, 29, 85, **89**, **145**, **42**, **20**, **4**, **16**, **37**, **58** (, **89**)]
9,992	[9,992, 247, 69, 117, 51, 26, 40, 16, 37, 58, **89**, **145**, **42**, **20**, **4**, **16**, **37**, **58** (, **89**)]
9,993	[9,993, 252, 33, 18, 65, 61, 37, 58, **89**, **145**, **42**, **20**, **4**, **16**, **37**, **58** (, **89**)]
9,994	[9,994, 259, 110, 2, 4, 16, 37, 58, **89**, **145**, **42**, **20**, **4**, **16**, **37**, **58** (, **89**)]
9,995	[9,995, 268, 104, 17, 50, 25, 29, 85, **89**, **145**, **42**, **20**, **4**, **16**, **37**, **58** (, **89**)]
9,996	[9,996, 279, 134, 26, 40, 16, 37, 58, **89**, **145**, **42**, **20**, **4**, **16**, **37**, **58** (, **89**)]
9,997	[9,997, 292, **89**, **145**, **42**, **20**, **4**, **16**, **37**, **58** (, **89**)]
9,998	[9,998, 307, 58, **89**, **145**, **42**, **20**, **4**, **16**, **37**, **58** (, **89**)]
9,999	[9,999, 324, 29, 85, **89**, **145**, **42**, **20**, **4**, **16**, **37**, **58** (, **89**)]
10,000	[10,000, 100, **1** (, **1**)]

MORE NUMBER LOOPS

We can extend this exploration of number loops by taking the sums of other powers of the digits of a number. These will lead to other interesting results. Let's continue now by selecting any number and then finding the sum of the cubes of the digits. Of course, for *any* number—that is, a randomly selected number—you may reach a number different from the one with which you started. If so, repeat this process with each succeeding number (derived from a sum of the cubed digits) until you get into a "loop." A loop can be easily recognized when you reach a number that you already reached earlier or even the one you started with. This should become clearer with an example. Let's begin with the (arbitrarily selected) number **352** and find the sum of the cubes of the digits.

The sum of the cubes of the digits of **352** is $3^3 + 5^3 + 2^3 = 27 + 125 + 8 = 160$.

Now we use this sum, 160, and repeat the process:

The sum of the cubes of the digits of **160** is $1^3 + 6^3 + 0^3 = 1 + 216 + 0 = 217$.

Again repeat the process with 217:

The sum of the cubes of the digits of **217** is $2^3 + 1^3 + 7^3 = 8 + 1 + 343 = 352$.

Surprise! This is the same number (**352**) we started with.

Not to take any of the suspense out of the joy of determining a loop, we will tell you now that there are only five numbers by which the loop is reached in one step. They are the following:

$$1 \rightarrow 1^3 \qquad\qquad\quad = 1$$
$$153 \rightarrow 1^3 + 5^3 + 3^3 = 1 + 125 + 27 \quad = 153$$
$$370 \rightarrow 3^3 + 7^3 + 0^3 = 27 + 343 + 0 \quad = 370$$
$$371 \rightarrow 3^3 + 7^3 + 1^3 = 27 + 343 + 1 \quad = 371$$
$$407 \rightarrow 4^3 + 0^3 + 7^3 = 64 + 0 + 343 \quad = 407$$

If we select, as a starting number, any number between 1 and 100, and we take the sum of the cubes of the digits of the number, we will eventually enter one of the following loops:

[1 (, 1)],
[55, 250, 133 (, 55)]   = [133, 55, 250 (, 133)] = [250, 133, 55 (, 250)],
[133, 55, 250 (, 133)]   = [250, 133, 55 (, 250)] = [55, 250, 133 (, 55)],
[153 (, 153)],
[160, 217, 352 (, 160)] = [217, 352, 160 (, 217)] = [352, 160, 217 (, 352)],
[250, 133, 55 (, 250)]   = [55, 250, 133 (, 55)] = [133, 55, 250 (, 133)],
[370 (, 370)],
[371 (, 371)],
[407 (, 407)],
[919, 1459 (, 919)]   = [1459, 919 (, 1459)]

If we select, as a starting number, one between 101 and 1,000, then we can enter an additional end loop:

[136, 244 (, 136)] = [244, 136 (, 244)]

You may now want to experience the fun of getting into these loops; so we offer the following table of the progression of the sums of the cubes of the digits for the numbers $n = 1, 2, ..., 100$.

n	Sequence of the sums of the digits cubed
1	[1 (, 1)]
2	[2, 8, 512, 134, 92, 737, 713, 371 (, 371)]
3	[3, 27, 351, 153 (, 153)]
4	[4, 64, 280, 520, 133, 55, 250 (, 133)]
5	[5, 125, 134, 92, 737, 713, 371 (, 371)]
6	[6, 216, 225, 141, 66, 432, 99, 1458, 702, 351, 153 (, 153)]
7	[7, 343, 118, 514, 190, 730, 370 (, 370)]
8	[8, 512, 134, 92, 737, 713, 371 (, 371)]
9	[9, 729, 1080, 513, 153 (, 153)]
10	[10, 1 (, 1)]
11	[11, 2, 8, 512, 134, 92, 737, 713, 371 (, 371)]
12	[12, 9, 729, 1080, 513, 153 (, 153)]
13	[13, 28, 520, 133, 55, 250 (, 133)]
14	[14, 65, 341, 92, 737, 713, 371 (, 371)]
15	[15, 126, 225, 141, 66, 432, 99, 1458, 702, 351, 153 (, 153)]
16	[16, 217, 352, 160 (, 217)]
17	[17, 344, 155, 251, 134, 92, 737, 713, 371 (, 371)]
18	[18, 513, 153 (, 153)]
19	[19, 730, 370 (, 370)]
20	[20, 8, 512, 134, 92, 737, 713, 371 (, 371)]
21	[21, 9, 729, 1080, 513, 153 (, 153)]
22	[22, 16, 217, 352, 160 (, 217)]
23	[23, 35, 152, 134, 92, 737, 713, 371 (, 371)]
24	[24, 72, 351, 153 (, 153)]
25	[25, 133, 55, 250 (, 133)]
26	[26, 224, 80, 512, 134, 92, 737, 713, 371 (, 371)]
27	[27, 351, 153 (, 153)]
28	[28, 520, 133, 55, 250 (, 133)]
29	[29, 737, 713, 371 (, 371)]
30	[30, 27, 351, 153 (, 153)]
31	[31, 28, 520, 133, 55, 250 (, 133)]
32	[32, 35, 152, 134, 92, 737, 713, 371 (, 371)]
33	[33, 54, 189, 1242, 81, 513, 153 (, 153)]
34	[34, 91, 730, 370 (, 370)]
35	[35, 152, 134, 92, 737, 713, 371 (, 371)]
36	[36, 243, 99, 1458, 702, 351, 153 (, 153)]
37	[37, 370 (, 370)]
38	[38, 539, 881, 1025, 134, 92, 737, 713, 371 (, 371)]
39	[39, 756, 684, 792, 1080, 513, 153 (, 153)]
40	[40, 64, 280, 520, 133, 55, 250 (, 133)]
41	[41, 65, 341, 92, 737, 713, 371 (, 371)]

n	Sequence of the sums of the digits cubed
42	[42, 72, 351, 153 (, 153)]
43	[43, 91, 730, 370 (, 370)]
44	[44, 128, 521, 134, 92, 737, 713, 371 (, 371)]
45	[45, 189, 1242, 81, 513, 153 (, 153)]
46	[46, 280, 520, 133, 55, 250 (, 133)]
47	[47, 407 (, 407)]
48	[48, 576, 684, 792, 1080, 513, 153 (, 153)]
49	[49, 793, 1099, 1459, 919 (, 1459)]
50	[50, 125, 134, 92, 737, 713, 371 (, 371)]
51	[51, 126, 225, 141, 66, 432, 99, 1458, 702, 351, 153 (, 153)]
52	[52, 133, 55, 250 (, 133)]
53	[53, 152, 134, 92, 737, 713, 371 (, 371)]
54	[54, 189, 1242, 81, 513, 153 (, 153)]
55	[55, 250, 133 (, 55)]
56	[56, 341, 92, 737, 713, 371 (, 371)]
57	[57, 468, 792, 1080, 513, 153 (, 153)]
58	[58, 637, 586, 853, 664, 496, 1009, 730, 370 (, 370)]
59	[59, 854, 701, 344, 155, 251, 134, 92, 737, 713, 371 (, 371)]
60	[60, 216, 225, 141, 66, 432, 99, 1458, 702, 351, 153 (, 153)]
61	[61, 217, 352, 160 (, 217)]
62	[62, 224, 80, 512, 134, 92, 737, 713, 371 (, 371)]
63	[63, 243, 99, 1458, 702, 351, 153 (, 153)]
64	[64, 280, 520, 133, 55, 250 (, 133)]
65	[65, 341, 92, 737, 713, 371 (, 371)]
66	[66, 432, 99, 1458, 702, 351, 153 (, 153)]
67	[67, 559, 979, 1801, 514, 190, 730, 370 (, 370)]
68	[68, 728, 863, 755, 593, 881, 1025, 134, 92, 737, 713, 371 (, 371)]
69	[69, 945, 918, 1242, 81, 513, 153 (, 153)]
70	[70, 343, 118, 514, 190, 730, 370 (, 370)]
71	[71, 344, 155, 251, 134, 92, 737, 713, 371 (, 371)]
72	[72, 351, 153 (, 153)]
73	[73, 370 (, 370)]
74	[74, 407 (, 407)]
75	[75, 468, 792, 1080, 513, 153 (, 153)]
76	[76, 559, 979, 1801, 514, 190, 730, 370 (, 370)]
77	[77, 686, 944, 857, 980, 1241, 74, 407 (, 407)]
78	[78, 855, 762, 567, 684, 792, 1080, 513, 153 (, 153)]
79	[79, 1072, 352, 160, 217 (, 352)]
80	[80, 512, 134, 92, 737, 713, 371 (, 371)]
81	[81, 513, 153 (, 153)]
82	[82, 520, 133, 55, 250 (, 133)]

n	Sequence of the sums of the digits cubed
83	[83, 539, 881, 1025, 134, 92, 737, 713, 371 (, 371)]
84	[84, 576, 684, 792, 1080, 513, 153 (, 153)]
85	[85, 637, 586, 853, 664, 496, 1009, 730, 370 (, 370)]
86	[86, 728, 863, 755, 593, 881, 1025, 134, 92, 737, 713, 371 (, 371)]
87	[87, 855, 762, 567, 684, 792, 1080, 513, 153 (, 153)]
88	[88, 1024, 73, 370 (, 370)]
89	[89, 1241, 74, 407 (, 407)]
90	[90, 729, 1080, 513, 153 (, 153)]
91	[91, 730, 370 (, 370)]
92	[92, 737, 713, 371 (, 371)]
93	[93, 756, 684, 792, 1080, 513, 153 (, 153)]
94	[94, 793, 1099, 1459, 919 (, 1459)]
95	[95, 854, 701, 344, 155, 251, 134, 92, 737, 713, 371 (, 371)]
96	[96, 945, 918, 1242, 81, 513, 153 (, 153)]
97	[97, 1072, 352, 160, 217 (, 352)]
98	[98, 1241, 74, 407 (, 407)]
99	[99, 1458, 702, 351, 153 (, 153)]
100	[100, 1 (, 1)]

You will also notice that for $n = 112$ and $n = 787$, the sums of the cubes of the digits eventually end up at the loop [1 (, 1)]:

$$[112, 10, 1 (, 1)]$$
$$[787, 1198, 1243, 100, 1 (, 1)]$$

We can continue this scheme with higher powers, such as for the fourth power. Consider the following:

$$\mathbf{1,138} \rightarrow 1^4 + 1^4 + 3^4 + 8^4 = 4,179$$
$$4,179 \rightarrow 4^4 + 1^4 + 7^4 + 9^4 = 9,219$$
$$9,219 \rightarrow 9^4 + 2^4 + 1^4 + 9^4 = 13,139$$
$$13,139 \rightarrow 1^4 + 3^4 + 1^4 + 3^4 + 9^4 = 6,725$$
$$6,725 \rightarrow 6^4 + 7^4 + 2^4 + 5^4 = 4,338$$
$$4,338 \rightarrow 4^4 + 3^4 + 3^4 + 8^4 = 4,514$$
$$4,514 \rightarrow 4^4 + 5^4 + 1^4 + 4^4 = \mathbf{1,138}$$

You may want to experiment with the sums of the powers of the digits of any number and see what interesting results it may lead to. Look for patterns of loops, and see if you can determine the extent of a loop based on the nature of the original number.

YET ANOTHER (FAMOUS) LOOP

Some might say that there is a quirk in our decimal number system that manifests itself with the following loop. There isn't much you can do with it, other than to marvel at the amazing outcome—that should be enough!

You may wish to use a calculator, unless you want to have practice in subtraction.

So, here is the process of this miraculous loop. Follow the steps below.

1. **Begin by selecting a four-digit number (except one that has all digits the same).**
2. **Rearrange the digits of the number so that they form the largest number possible. (That means write the number with the digits in descending order.)**
3. **Then rearrange the digits of the number so that they form the smallest number possible. (That means write the number with the digits in ascending order. Zeros can take the first few places.)**
4. **Subtract these two numbers (obviously, the smaller from the larger).**
5. **Take this difference and continue the process, over and over and over, until you notice something disturbing happening. Don't give up before something unusual happens.**

Depending on the number you selected to start with, you will eventually arrive at the number **6,174**—perhaps after one subtraction, or after several subtractions. When you do, you will find yourself in an endless loop. Remember that you began with an arbitrarily selected number. Although some readers might be motivated to investigate this further, others will just sit back in awe.

Here is an example of how this works with our arbitrarily selected starting number 3,203.

We will select the number   3,203.
The *largest* number formed with these digits is   3,320
The *smallest* number formed with these digits is   0,233
The difference is   3,087

Now using this number, 3,087, we continue the process:

The largest number formed with these digits is   8,730
The smallest number formed with these digits is   0,378
The difference is   8,352

Again, we repeat the process:

> The largest number formed with these digits is   8,532
> The smallest number formed with these digits is   2,358
> The difference is   6,174
>
> The largest number formed with these digits is   7,641
> The smallest number formed with these digits is   1,467
> The difference is   6,174

This nifty loop was first discovered by the Indian mathematician Dattathreya Ramachandra Kaprekar (1905–1986) in 1946.[2] We often refer to the number 6,174 as the *Kaprekar Constant.*[3]

And so the loop is formed, since once you arrive at 6,174, you keep on getting back to 6,174. Remember, all this began with an arbitrarily selected **four-digit** number and will always end up with the number 6,174, which then gets you into an endless loop (i.e., we continuously get back to 6,174).

These examples of loops are easy to verify since there are a finite number of four-digit numbers.

For the experts, or the curious, more can be found about this number.[4]

Here are some variations for you to ponder—and appreciate!

SOME VARIATIONS OF THE KAPREKAR CONSTANTS

- If you choose a two-digit number (not one with two of the same digits), then the Kaprekar Constant would be 81 and you would end up in a loop of length 5: [81, 63, 27, 45, 09 (,81)]. There is no loop of length 1 for two-digit numbers, as we had earlier.

---

2. Kaprekar announced it at the Madras Mathematical Conference in 1949. He published the result in the paper "Problems Involving Reversal of Digits," in *Scripta Mathematica* in 1953; see also D. R. Kaprekar, "An Interesting Property of the Number 6174," *Scripta Mathematica* 15 (1955): 244–45.

3. We should not confuse this number (6,174) with other *Kaprekar numbers*: 9, 45, 297, 703, 4,879, ...:

$9^2 = 81$, $8 + 1 = 9$;
$45^2 = 2,025$, $20 + 25 = 45$;
$297^2 = 88,209$, $88 + 209 = 297$;
$703^2 = 494,209$, $494 + 209 = 703$;
$4,879^2 = 23,804,641$, $238 + 4,641 = 4,879$;
$17,344^2 = 300,814,336$, $3,008 + 14,336 = 17,344$;
$538,461^2 = 289,940,248,521$, $289,940 + 248,521 = 538,461$

4.   Malcolm E. Lines, *A Number for Your Thoughts: Facts and Speculations about Numbers* (Bristol, UK: Hilger, 1986); Robert W. Ellis and Jason R. Lewis, "Investigations into the Kaprekar Process," http://www.rose-hulman.edu/mathjournal/archives/2002/vol3-n2/paper4/v3n2-4pd.pdf.

- If you choose a three-digit number (not one with all of the same digits), then the Kaprekar Constant would be 495 and you would end up in a loop of length 1: [495 (, 495)].

- If you choose a four-digit number (not one with all of the same digits), then the Kaprekar Constant would be 6,174—as we have seen before—and you end up with a loop of length 1: [6174 (, 6174)].

- If you choose a five-digit number (not one of all same digits), then there are three Kaprekar Constants: 53,955; 61,974; and 62,964.

  One of length 2   [53,955; 59,994 (; 53,955)],
  and two of length 4   [61,974; 82,962; 75,933; 63,954 (; 61,974)]
                                 [62,964; 71,973; 83,952; 74,943 (; 62,964)]

You can follow this scheme with six-digit numbers, and you will also find yourself getting into a loop. One number you may find leading you into the loop is 840,852, but do not let this stop you from further investigating this mathematical curiosity.[5] For example, consider the digit sum of each difference. Since the sum of the digits of the subtrahend and the minuend[6] are the same, the difference will have a digit sum that is a multiple of 9. For three- and four-digit numbers, the digit sum is 18. In the case of five- and six-digit numbers, the digit sum appears to be 27. It follows that for seven- and eight-digit numbers, the digit sum is 36. Yes, you will find that the digit sum, when this technique is used on nine- and ten-digit number, is 45. You will be pleasantly surprised when you check to see what the digit sum is for even larger numbers.

THE ULAM-COLLATZ LOOP

There are times when we think of the beauty in nature as magical. Is nature magical? Some feel that when something is truly surprising and "neat," it is beautiful. From that standpoint, we will show a seemingly "magical" property in mathematics. This is one that has baffled mathematicians for many years and still no one knows why it happens. Try it, you'll like it.

---

5. If you choose a six-digit number (not one with all of the same digits), then there are also three Kaprekar Constants: 549,945; 631,764; and 420,876. Two of length 1: [549,945 (;549,945)]; [631,764 (; 631,764)] and one of length 7: [420,876; 851,742; 750,843; 840,852; 860,832; 862,632; 642,654 (; 420,876)].
If you choose a seven-digit number (not one with all of the same digits), then there is only one Kaprekar constant: 7,509,843.
There is a loop of length 8: [7,509,843; 9,529,641; 8,719,722; 8,649,432; 7,519,743; 8,429,652; 7,619,733; 8,439,552 (; 7,509,843)].
6. In a subtraction, the number in the *subtrahend* is subtracted from the number in the *minuend* to get the result, referred to as the *difference*.

We begin by asking you to follow two rules as you work with any *arbitrarily* selected number.

If the number is *odd*, then multiply by 3 and add 1.

If the number is *even*, then divide by 2.

Regardless of the number you select, after continued repetition of the process, you will always end up with the number 1.

Let's try it for the *arbitrarily selected* number **7**:

7 is odd, so we multiply by 3 and add 1 to get **22**.

22 is even, so we simply divide by 2 to get **11**.

11 is odd, so we multiply by 3 and add 1 to get **34**.

34 is even, so we divide by 2 to get **17**.

17 is odd, so we multiply by 3 and add 1 to get **52**.

52 is even, so we divide by 2 to get **26**.

26 is even, so we divide by 2 to get **13**.

13 is odd, so we multiply by 3 and add 1 to get **40**.

40 is even, so we divide by 2 to get **20**.

20 is even, so we divide by 2 to get **10**.

10 is even, so we divide by 2 to get **5**.

5 is odd, so we multiply by 3 and add 1 to get **16**.

16 is even, so we divide by 2 to get **8**.

8 is even, so we divide by 2 to get **4**.

4 is even, so we divide by 2 to get **2**.

2 is also even, so we again divide by 2 to get **1**.

If we were to continue, we would find ourselves in a loop.

[1 is odd, so we multiply by 3 and add 1 to get **4**, …]. After sixteen steps, we end up with a 1, that, if we continue the process, would lead us back to 4, and then onto the 1 again. We hit a loop!

Therefore, we get the sequence 7, 22, 11, 34, 17, 52, 26, 13, 40, 20, 10, 5, 16, 8, **4, 2, 1, 4, 2, 1** , . . .

The following schematic (figure 3-6) will show the path we have just taken:

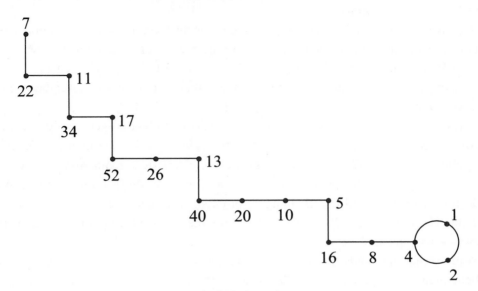

Figure 3-6 *n* = 7

It is also interesting to see a graph (figure 3-7) of the steps of this process.

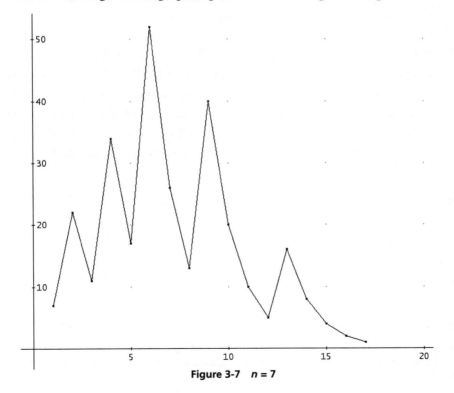

Figure 3-7 *n* = 7

No matter which number we begin with (here we started with **7**), we will eventually get to 1.

This is truly remarkable! Try it for some other numbers to convince yourself that it really does work. Had we started with **9** as our arbitrarily selected number, it would have required **19** steps to reach 1. Starting with **41** will require 109 steps to reach 1.

One of the amazing features of this scheme is that no matter which number we begin with, we always end up with the following loop of length 3: [4, 2, 1 (, 4)].

Does this *really* work for *all* numbers? This amazing little loop-generating scheme was first discovered in 1932 by Lothar Collatz (1910–1990), a German mathematician, who then published it in 1937. Credit is also given to the American mathematician Stanislaus Marcin Ulam (1909–1984), who worked on the Manhattan Project during World War II. Credit for the discovery of this challenging problem is further given to the German mathematician Helmut Hasse (1898–1979). Therefore, the scheme (or algorithm) can be found under various names.

A proof that this holds for all numbers has not yet been found. The famous Canadian mathematician H. S. M. Coxeter (1907–2003) offered a prize of $50 to anyone who could come up with such a proof, and $100 for anyone who could find a number for which this doesn't work. Later, the Hungarian mathematician Paul Erdös (1913–1996) raised the prize money to $500. Still, with all these and many further incentives, no one has yet found a proof. This seemingly "true" algorithm then must remain a conjecture until it is proved true for all cases.

Most recently (with aid of computers), this "$3n + 1$ problem," as it is also commonly known, has been shown to be true for the numbers up to $18 \cdot 2^{58} \approx 5.188146770 \cdot 10^{18}$ (June 1, 2008)[7]—that means, for more than 5 quintillion [in Europe: trillion] it is proved.

You may also like to see a graphic "picture" of the path of this curious number property. We offer you a schematic that shows the sequence of start numbers 1–20. In figure 3-8, the bold numbers (1–20) can be starting points for your progression following the above rules.

---

7. See http://www.ieeta.pt/~tos/3x+1.html. Also see Tomás Oliveira e Silva, "Maximum Excursion and Stopping Time Record-Holders for the $3x + 1$ Problem: Computational Results," *Mathematics of Computation* 68, no. 225 (1999): 371–84.

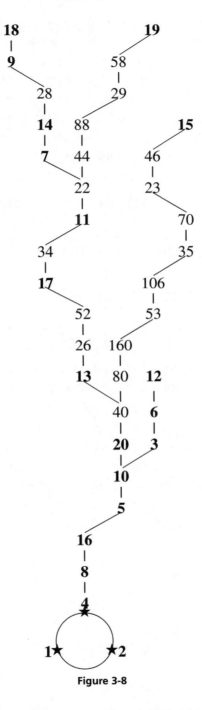

Figure 3-8

For beginning numbers 11, 34, 35, 96, 104, 106, 113, 320, 336, 340, 341, and 1,024, we get the following graphic representation (figure 3-9):

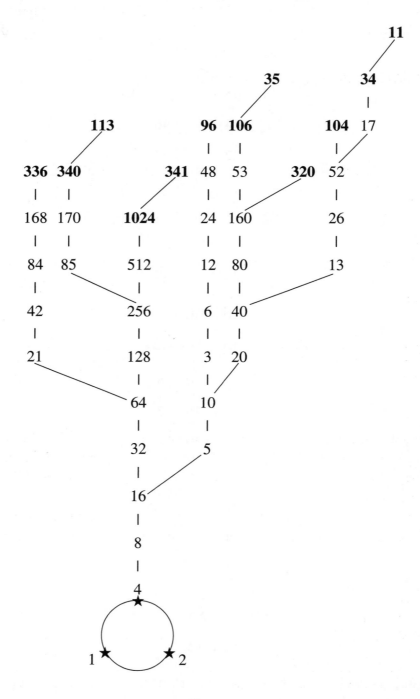

Figure 3-9

In this way the numbers determine a close relationship even if they are far apart in their relative sizes (e.g., 3 and 20; 80 and 13). You can see this in a larger sense with the following (figure 3-10):

Ulam-Collatz Sequences for *n* = 1, 2, ... , 25

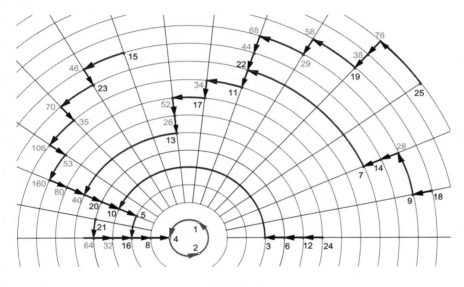

Figure 3-10

Notice that you will always end up with the final loop of 4-2-1. That is, when you reach 4, you will always get to the 1, and then, were you to try to continue, after having arrived at the 1, you will always get back to the 1, since, by applying the "3*n*+1 rule" [3·1+1 = 4], you again enter the loop 4-2-1.

To provide you with some guidance (and perhaps some encouragement as well), we offer you a chart that shows the steps that each number *n* will require in order to reach the 1.

Ulam-Collatz Sequences for *n* = 0, 1, 2, ... , 100

n	Ulam-Collatz Sequence
0	[0, 0]
1	[1, 4, 2, 1] = [1, **4, 2, 1** (, 4)]
2	[2, 1, 4, 2] = [2, 1, **4, 2, 1** (, 4)]
3	[3, 10, 5, 16, 8, **4, 2, 1** (, 4)]
4	[**4, 2, 1** (, 4)]

n	Ulam-Collatz Sequence
5	[5, 16, 8, **4**, **2**, **1** (, 4)]
6	[6, 3, 10, 5, 16, 8, **4**, **2**, **1** (, 4)]
7	[7, 22, 11, 34, 17, 52, 26, 13, 40, 20, 10, 5, 16, 8, **4**, **2**, **1** (, 4)]
8	[8, **4**, **2**, **1** (, 4)]
9	[9, 28, 14, 7, 22, 11, 34, 17, 52, 26, 13, 40, 20, 10, 5, 16, 8, **4**, **2**, **1** (, 4)]
10	[10, 5, 16, 8, **4**, **2**, **1** (, 4)]
11	[11, 34, 17, 52, 26, 13, 40, 20, 10, 5, 16, 8, **4**, **2**, **1** (, 4)]
12	[12, 6, 3, 10, 5, 16, 8, **4**, **2**, **1** (, 4)]
13	[13, 40, 20, 10, 5, 16, 8, **4**, **2**, **1** (, 4)]
14	[14, 7, 22, 11, 34, 17, 52, 26, 13, 40, 20, 10, 5, 16, 8, **4**, **2**, **1** (, 4)]
15	[15, 46, 23, 70, 35, 106, 53, 160, 80, 40, 20, 10, 5, 16, 8, **4**, **2**, **1** (, 4)]
16	[16, 8, **4**, **2**, **1** (, 4)]
17	[17, 52, 26, 13, 40, 20, 10, 5, 16, 8, **4**, **2**, **1** (, 4)]
18	[18, 9, 28, 14, 7, 22, 11, 34, 17, 52, 26, 13, 40, 20, 10, 5, 16, 8, **4**, **2**, **1** (, 4)]
19	[19, 58, 29, 88, 44, 22, 11, 34, 17, 52, 26, 13, 40, 20, 10, 5, 16, 8, **4**, **2**, **1** (, 4)]
20	[20, 10, 5, 16, 8, **4**, **2**, **1** (, 4)]
21	[21, 64, 32, 16, 8, **4**, **2**, **1** (, 4)]
22	[22, 11, 34, 17, 52, 26, 13, 40, 20, 10, 5, 16, 8, **4**, **2**, **1** (, 4)]
23	[23, 70, 35, 106, 53, 160, 80, 40, 20, 10, 5, 16, 8, **4**, **2**, **1** (, 4)]
24	[24, 12, 6, 3, 10, 5, 16, 8, **4**, **2**, **1** (, 4)]
25	[25, 76, 38, 19, 58, 29, 88, 44, 22, 11, 34, 17, 52, 26, 13, 40, 20, 10, 5, 16, 8, **4**, **2**, **1** (, 4)]
26	[26, 13, 40, 20, 10, 5, 16, 8, **4**, **2**, **1** (, 4)]
27	[27, 82, 41, 124, 62, 31, 94, 47, 142, 71, 214, 107, 322, 161, 484, 242, 121, 364, 182, 91, 274, 137, 412, 206, 103, 310, 155, 466, 233, 700, 350, 175, 526, 263, 790, 395, 1186, 593, 1780, 890, 445, 1336, 668, 334, 167, 502, 251, 754, 377, 1132, 566, 283, 850, 425, 1276, 638, 319, 958, 479, 1438, 719, 2158, 1079, 3238, 1619, 4858, 2429, 7288, 3644, 1822, 911, 2734, 1367, 4102, 2051, 6154, 3077, 9232, 4616, 2308, 1154, 577, 1732, 866, 433, 1300, 650, 325, 976, 488, 244, 122, 61, 184, 92, 46, 23, 70, 35, 106, 53, 160, 80, 40, 20, 10, 5, 16, 8, **4**, **2**, **1** (, 4)]
28	[28, 14, 7, 22, 11, 34, 17, 52, 26, 13, 40, 20, 10, 5, 16, 8, **4**, **2**, **1** (, 4)]
29	[29, 88, 44, 22, 11, 34, 17, 52, 26, 13, 40, 20, 10, 5, 16, 8, **4**, **2**, **1** (, 4)]
30	[30, 15, 46, 23, 70, 35, 106, 53, 160, 80, 40, 20, 10, 5, 16, 8, **4**, **2**, **1** (, 4)]
31	[31, 94, 47, 142, 71, 214, 107, 322, 161, 484, 242, 121, 364, 182, 91, 274, 137, 412, 206, 103, 310, 155, 466, 233, 700, 350, 175, 526, 263, 790, 395, 1186, 593, 1780, 890, 445, 1336, 668, 334, 167, 502, 251, 754, 377, 1132, 566, 283, 850, 425, 1276, 638, 319, 958, 479, 1438, 719, 2158, 1079, 3238, 1619, 4858, 2429, 7288, 3644, 1822, 911, 2734, 1367, 4102, 2051, 6154, 3077, 9232, 4616, 2308, 1154, 577, 1732, 866, 433, 1300, 650, 325, 976, 488, 244, 122, 61, 184, 92, 46, 23, 70, 35, 106, 53, 160, 80, 40, 20, 10, 5, 16, 8, **4**, **2**, **1** (, 4)]
32	[32, 16, 8, **4**, **2**, **1** (, 4)]
33	[33, 100, 50, 25, 76, 38, 19, 58, 29, 88, 44, 22, 11, 34, 17, 52, 26, 13, 40, 20, 10, 5, 16, 8, **4**, **2**, **1** (, 4)]
34	[34, 17, 52, 26, 13, 40, 20, 10, 5, 16, 8, **4**, **2**, **1** (, 4)]
35	[35, 106, 53, 160, 80, 40, 20, 10, 5, 16, 8, **4**, **2**, **1** (, 4)]
36	[36, 18, 9, 28, 14, 7, 22, 11, 34, 17, 52, 26, 13, 40, 20, 10, 5, 16, 8, **4**, **2**, **1** (, 4)]
37	[37, 112, 56, 28, 14, 7, 22, 11, 34, 17, 52, 26, 13, 40, 20, 10, 5, 16, 8, **4**, **2**, **1** (, 4)]
38	[38, 19, 58, 29, 88, 44, 22, 11, 34, 17, 52, 26, 13, 40, 20, 10, 5, 16, 8, **4**, **2**, **1** (, 4)]

n	Ulam-Collatz Sequence
39	[39, 118, 59, 178, 89, 268, 134, 67, 202, 101, 304, 152, 76, 38, 19, 58, 29, 88, 44, 22, 11, 34, 17, 52, 26, 13, 40, 20, 10, 5, 16, 8, **4**, **2**, **1** (, 4)]
40	[40, 20, 10, 5, 16, 8, **4**, **2**, **1** (, 4)]
41	[41, 124, 62, 31, 94, 47, 142, 71, 214, 107, 322, 161, 484, 242, 121, 364, 182, 91, 274, 137, 412, 206, 103, 310, 155, 466, 233, 700, 350, 175, 526, 263, 790, 395, 1186, 593, 1780, 890, 445, 1336, 668, 334, 167, 502, 251, 754, 377, 1132, 566, 283, 850, 425, 1276, 638, 319, 958, 479, 1438, 719, 2158, 1079, 3238, 1619, 4858, 2429, 7288, 3644, 1822, 911, 2734, 1367, 4102, 2051, 6154, 3077, 9232, 4616, 2308, 1154, 577, 1732, 866, 433, 1300, 650, 325, 976, 488, 244, 122, 61, 184, 92, 46, 23, 70, 35, 106, 53, 160, 80, 40, 20, 10, 5, 16, 8, **4**, **2**, **1** (, 4)]
42	[42, 21, 64, 32, 16, 8, **4**, **2**, **1** (, 4)]
43	[43, 130, 65, 196, 98, 49, 148, 74, 37, 112, 56, 28, 14, 7, 22, 11, 34, 17, 52, 26, 13, 40, 20, 10, 5, 16, 8, **4**, **2**, **1** (, 4)]
44	[44, 22, 11, 34, 17, 52, 26, 13, 40, 20, 10, 5, 16, 8, **4**, **2**, **1** (, 4)]
45	[45, 136, 68, 34, 17, 52, 26, 13, 40, 20, 10, 5, 16, 8, **4**, **2**, **1** (, 4)]
46	[46, 23, 70, 35, 106, 53, 160, 80, 40, 20, 10, 5, 16, 8, **4**, **2**, **1** (, 4)]
47	[47, 142, 71, 214, 107, 322, 161, 484, 242, 121, 364, 182, 91, 274, 137, 412, 206, 103, 310, 155, 466, 233, 700, 350, 175, 526, 263, 790, 395, 1186, 593, 1780, 890, 445, 1336, 668, 334, 167, 502, 251, 754, 377, 1132, 566, 283, 850, 425, 1276, 638, 319, 958, 479, 1438, 719, 2158, 1079, 3238, 1619, 4858, 2429, 7288, 3644, 1822, 911, 2734, 1367, 4102, 2051, 6154, 3077, 9232, 4616, 2308, 1154, 577, 1732, 866, 433, 1300, 650, 325, 976, 488, 244, 122, 61, 184, 92, 46, 23, 70, 35, 106, 53, 160, 80, 40, 20, 10, 5, 16, 8, **4**, **2**, **1** (, 4)]
48	[48, 24, 12, 6, 3, 10, 5, 16, 8, **4**, **2**, **1** (, 4)]
49	[49, 148, 74, 37, 112, 56, 28, 14, 7, 22, 11, 34, 17, 52, 26, 13, 40, 20, 10, 5, 16, 8, **4**, **2**, **1** (, 4)]
50	[50, 25, 76, 38, 19, 58, 29, 88, 44, 22, 11, 34, 17, 52, 26, 13, 40, 20, 10, 5, 16, 8, **4**, **2**, **1** (, 4)]
51	[51, 154, 77, 232, 116, 58, 29, 88, 44, 22, 11, 34, 17, 52, 26, 13, 40, 20, 10, 5, 16, 8, **4**, **2**, **1** (, 4)]
52	[52, 26, 13, 40, 20, 10, 5, 16, 8, **4**, **2**, **1** (, 4)]
53	[53, 160, 80, 40, 20, 10, 5, 16, 8, **4**, **2**, **1** (, 4)]
54	[54, 27, 82, 41, 124, 62, 31, 94, 47, 142, 71, 214, 107, 322, 161, 484, 242, 121, 364, 182, 91, 274, 137, 412, 206, 103, 310, 155, 466, 233, 700, 350, 175, 526, 263, 790, 395, 1186, 593, 1780, 890, 445, 1336, 668, 334, 167, 502, 251, 754, 377, 1132, 566, 283, 850, 425, 1276, 638, 319, 958, 479, 1438, 719, 2158, 1079, 3238, 1619, 4858, 2429, 7288, 3644, 1822, 911, 2734, 1367, 4102, 2051, 6154, 3077, 9232, 4616, 2308, 1154, 577, 1732, 866, 433, 1300, 650, 325, 976, 488, 244, 122, 61, 184, 92, 46, 23, 70, 35, 106, 53, 160, 80, 40, 20, 10, 5, 16, 8, **4**, **2**, **1** (, 4)]
55	[55, 166, 83, 250, 125, 376, 188, 94, 47, 142, 71, 214, 107, 322, 161, 484, 242, 121, 364, 182, 91, 274, 137, 412, 206, 103, 310, 155, 466, 233, 700, 350, 175, 526, 263, 790, 395, 1186, 593, 1780, 890, 445, 1336, 668, 334, 167, 502, 251, 754, 377, 1132, 566, 283, 850, 425, 1276, 638, 319, 958, 479, 1438, 719, 2158, 1079, 3238, 1619, 4858, 2429, 7288, 3644, 1822, 911, 2734, 1367, 4102, 2051, 6154, 3077, 9232, 4616, 2308, 1154, 577, 1732, 866, 433, 1300, 650, 325, 976, 488, 244, 122, 61, 184, 92, 46, 23, 70, 35, 106, 53, 160, 80, 40, 20, 10, 5, 16, 8, **4**, **2**, **1** (, 4)]
56	[56, 28, 14, 7, 22, 11, 34, 17, 52, 26, 13, 40, 20, 10, 5, 16, 8, **4**, **2**, **1** (, 4)]
57	[57, 172, 86, 43, 130, 65, 196, 98, 49, 148, 74, 37, 112, 56, 28, 14, 7, 22, 11, 34, 17, 52, 26, 13, 40, 20, 10, 5, 16, 8, **4**, **2**, **1** (, 4)]

n	Ulam-Collatz Sequence
58	[58, 29, 88, 44, 22, 11, 34, 17, 52, 26, 13, 40, 20, 10, 5, 16, 8, **4, 2, 1** (, 4)]
59	[59, 178, 89, 268, 134, 67, 202, 101, 304, 152, 76, 38, 19, 58, 29, 88, 44, 22, 11, 34, 17, 52, 26, 13, 40, 20, 10, 5, 16, 8, **4, 2, 1** (, 4)]
60	[60, 30, 15, 46, 23, 70, 35, 106, 53, 160, 80, 40, 20, 10, 5, 16, 8, **4, 2, 1** (, 4)]
61	[61, 184, 92, 46, 23, 70, 35, 106, 53, 160, 80, 40, 20, 10, 5, 16, 8, **4, 2, 1** (, 4)]
62	[62, 31, 94, 47, 142, 71, 214, 107, 322, 161, 484, 242, 121, 364, 182, 91, 274, 137, 412, 206, 103, 310, 155, 466, 233, 700, 350, 175, 526, 263, 790, 395, 1186, 593, 1780, 890, 445, 1336, 668, 334, 167, 502, 251, 754, 377, 1132, 566, 283, 850, 425, 1276, 638, 319, 958, 479, 1438, 719, 2158, 1079, 3238, 1619, 4858, 2429, 7288, 3644, 1822, 911, 2734, 1367, 4102, 2051, 6154, 3077, 9232, 4616, 2308, 1154, 577, 1732, 866, 433, 1300, 650, 325, 976, 488, 244, 122, 61, 184, 92, 46, 23, 70, 35, 106, 53, 160, 80, 40, 20, 10, 5, 16, 8, **4, 2, 1** (, 4)]
63	[63, 190, 95, 286, 143, 430, 215, 646, 323, 970, 485, 1456, 728, 364, 182, 91, 274, 137, 412, 206, 103, 310, 155, 466, 233, 700, 350, 175, 526, 263, 790, 395, 1186, 593, 1780, 890, 445, 1336, 668, 334, 167, 502, 251, 754, 377, 1132, 566, 283, 850, 425, 1276, 638, 319, 958, 479, 1438, 719, 2158, 1079, 3238, 1619, 4858, 2429, 7288, 3644, 1822, 911, 2734, 1367, 4102, 2051, 6154, 3077, 9232, 4616, 2308, 1154, 577, 1732, 866, 433, 1300, 650, 325, 976, 488, 244, 122, 61, 184, 92, 46, 23, 70, 35, 106, 53, 160, 80, 40, 20, 10, 5, 16, 8, **4, 2, 1** (, 4)]
64	[64, 32, 16, 8, **4, 2, 1** (, 4)]
65	[65, 196, 98, 49, 148, 74, 37, 112, 56, 28, 14, 7, 22, 11, 34, 17, 52, 26, 13, 40, 20, 10, 5, 16, 8, **4, 2, 1** (, 4)]
66	[66, 33, 100, 50, 25, 76, 38, 19, 58, 29, 88, 44, 22, 11, 34, 17, 52, 26, 13, 40, 20, 10, 5, 16, 8, **4, 2, 1** (, 4)]
67	[67, 202, 101, 304, 152, 76, 38, 19, 58, 29, 88, 44, 22, 11, 34, 17, 52, 26, 13, 40, 20, 10, 5, 16, 8, **4, 2, 1** (, 4)]
68	[68, 34, 17, 52, 26, 13, 40, 20, 10, 5, 16, 8, **4, 2, 1** (, 4)]
69	[69, 208, 104, 52, 26, 13, 40, 20, 10, 5, 16, 8, **4, 2, 1** (, 4)]
70	[70, 35, 106, 53, 160, 80, 40, 20, 10, 5, 16, 8, **4, 2, 1** (, 4)]
71	[71, 214, 107, 322, 161, 484, 242, 121, 364, 182, 91, 274, 137, 412, 206, 103, 310, 155, 466, 233, 700, 350, 175, 526, 263, 790, 395, 1186, 593, 1780, 890, 445, 1336, 668, 334, 167, 502, 251, 754, 377, 1132, 566, 283, 850, 425, 1276, 638, 319, 958, 479, 1438, 719, 2158, 1079, 3238, 1619, 4858, 2429, 7288, 3644, 1822, 911, 2734, 1367, 4102, 2051, 6154, 3077, 9232, 4616, 2308, 1154, 577, 1732, 866, 433, 1300, 650, 325, 976, 488, 244, 122, 61, 184, 92, 46, 23, 70, 35, 106, 53, 160, 80, 40, 20, 10, 5, 16, 8, **4, 2, 1** (, 4)]
72	[72, 36, 18, 9, 28, 14, 7, 22, 11, 34, 17, 52, 26, 13, 40, 20, 10, 5, 16, 8, **4, 2, 1** (, 4)]
73	[73, 220, 110, 55, 166, 83, 250, 125, 376, 188, 94, 47, 142, 71, 214, 107, 322, 161, 484, 242, 121, 364, 182, 91, 274, 137, 412, 206, 103, 310, 155, 466, 233, 700, 350, 175, 526, 263, 790, 395, 1186, 593, 1780, 890, 445, 1336, 668, 334, 167, 502, 251, 754, 377, 1132, 566, 283, 850, 425, 1276, 638, 319, 958, 479, 1438, 719, 2158, 1079, 3238, 1619, 4858, 2429, 7288, 3644, 1822, 911, 2734, 1367, 4102, 2051, 6154, 3077, 9232, 4616, 2308, 1154, 577, 1732, 866, 433, 1300, 650, 325, 976, 488, 244, 122, 61, 184, 92, 46, 23, 70, 35, 106, 53, 160, 80, 40, 20, 10, 5, 16, 8, **4, 2, 1** (, 4)]
74	[74, 37, 112, 56, 28, 14, 7, 22, 11, 34, 17, 52, 26, 13, 40, 20, 10, 5, 16, 8, **4, 2, 1** (, 4)]
75	[75, 226, 113, 340, 170, 85, 256, 128, 64, 32, 16, 8, **4, 2, 1** (, 4)]
76	[76, 38, 19, 58, 29, 88, 44, 22, 11, 34, 17, 52, 26, 13, 40, 20, 10, 5, 16, 8, **4, 2, 1** (, 4)]
77	[77, 232, 116, 58, 29, 88, 44, 22, 11, 34, 17, 52, 26, 13, 40, 20, 10, 5, 16, 8, **4, 2, 1** (, 4)]
78	[78, 39, 118, 59, 178, 89, 268, 134, 67, 202, 101, 304, 152, 76, 38, 19, 58, 29, 88, 44, 22, 11, 34, 17, 52, 26, 13, 40, 20, 10, 5, 16, 8, **4, 2, 1** (, 4)]

n	Ulam-Collatz Sequence
79	[79, 238, 119, 358, 179, 538, 269, 808, 404, 202, 101, 304, 152, 76, 38, 19, 58, 29, 88, 44, 22, 11, 34, 17, 52, 26, 13, 40, 20, 10, 5, 16, 8, **4, 2, 1** (, 4)]
80	[80, 40, 20, 10, 5, 16, 8, **4, 2, 1** (, 4)]
81	[81, 244, 122, 61, 184, 92, 46, 23, 70, 35, 106, 53, 160, 80, 40, 20, 10, 5, 16, 8, **4, 2, 1** (, 4)]
82	[82, 41, 124, 62, 31, 94, 47, 142, 71, 214, 107, 322, 161, 484, 242, 121, 364, 182, 91, 274, 137, 412, 206, 103, 310, 155, 466, 233, 700, 350, 175, 526, 263, 790, 395, 1186, 593, 1780, 890, 445, 1336, 668, 334, 167, 502, 251, 754, 377, 1132, 566, 283, 850, 425, 1276, 638, 319, 958, 479, 1438, 719, 2158, 1079, 3238, 1619, 4858, 2429, 7288, 3644, 1822, 911, 2734, 1367, 4102, 2051, 6154, 3077, 9232, 4616, 2308, 1154, 577, 1732, 866, 433, 1300, 650, 325, 976, 488, 244, 122, 61, 184, 92, 46, 23, 70, 35, 106, 53, 160, 80, 40, 20, 10, 5, 16, 8, **4, 2, 1** (, 4)]
83	[83, 250, 125, 376, 188, 94, 47, 142, 71, 214, 107, 322, 161, 484, 242, 121, 364, 182, 91, 274, 137, 412, 206, 103, 310, 155, 466, 233, 700, 350, 175, 526, 263, 790, 395, 1186, 593, 1780, 890, 445, 1336, 668, 334, 167, 502, 251, 754, 377, 1132, 566, 283, 850, 425, 1276, 638, 319, 958, 479, 1438, 719, 2158, 1079, 3238, 1619, 4858, 2429, 7288, 3644, 1822, 911, 2734, 1367, 4102, 2051, 6154, 3077, 9232, 4616, 2308, 1154, 577, 1732, 866, 433, 1300, 650, 325, 976, 488, 244, 122, 61, 184, 92, 46, 23, 70, 35, 106, 53, 160, 80, 40, 20, 10, 5, 16, 8, **4, 2, 1** (, 4)]
84	[84, 42, 21, 64, 32, 16, 8, **4, 2, 1** (, 4)]
85	[85, 256, 128, 64, 32, 16, 8, **4, 2, 1** (, 4)]
86	[86, 43, 130, 65, 196, 98, 49, 148, 74, 37, 112, 56, 28, 14, 7, 22, 11, 34, 17, 52, 26, 13, 40, 20, 10, 5, 16, 8, **4, 2, 1** (, 4)]
87	[87, 262, 131, 394, 197, 592, 296, 148, 74, 37, 112, 56, 28, 14, 7, 22, 11, 34, 17, 52, 26, 13, 40, 20, 10, 5, 16, 8, **4, 2, 1** (, 4)]
88	[88, 44, 22, 11, 34, 17, 52, 26, 13, 40, 20, 10, 5, 16, 8, **4, 2, 1** (, 4)]
89	[89, 268, 134, 67, 202, 101, 304, 152, 76, 38, 19, 58, 29, 88, 44, 22, 11, 34, 17, 52, 26, 13, 40, 20, 10, 5, 16, 8, **4, 2, 1** (, 4)]
90	[90, 45, 136, 68, 34, 17, 52, 26, 13, 40, 20, 10, 5, 16, 8, **4, 2, 1** (, 4)]
91	[91, 274, 137, 412, 206, 103, 310, 155, 466, 233, 700, 350, 175, 526, 263, 790, 395, 1186, 593, 1780, 890, 445, 1336, 668, 334, 167, 502, 251, 754, 377, 1132, 566, 283, 850, 425, 1276, 638, 319, 958, 479, 1438, 719, 2158, 1079, 3238, 1619, 4858, 2429, 7288, 3644, 1822, 911, 2734, 1367, 4102, 2051, 6154, 3077, 9232, 4616, 2308, 1154, 577, 1732, 866, 433, 1300, 650, 325, 976, 488, 244, 122, 61, 184, 92, 46, 23, 70, 35, 106, 53, 160, 80, 40, 20, 10, 5, 16, 8, **4, 2, 1** (, 4)]
92	[92, 46, 23, 70, 35, 106, 53, 160, 80, 40, 20, 10, 5, 16, 8, **4, 2, 1** (, 4)]
93	[93, 280, 140, 70, 35, 106, 53, 160, 80, 40, 20, 10, 5, 16, 8, **4, 2, 1** (, 4)]
94	[94, 47, 142, 71, 214, 107, 322, 161, 484, 242, 121, 364, 182, 91, 274, 137, 412, 206, 103, 310, 155, 466, 233, 700, 350, 175, 526, 263, 790, 395, 1186, 593, 1780, 890, 445, 1336, 668, 334, 167, 502, 251, 754, 377, 1132, 566, 283, 850, 425, 1276, 638, 319, 958, 479, 1438, 719, 2158, 1079, 3238, 1619, 4858, 2429, 7288, 3644, 1822, 911, 2734, 1367, 4102, 2051, 6154, 3077, 9232, 4616, 2308, 1154, 577, 1732, 866, 433, 1300, 650, 325, 976, 488, 244, 122, 61, 184, 92, 46, 23, 70, 35, 106, 53, 160, 80, 40, 20, 10, 5, 16, 8, **4, 2, 1** (, 4)]

n	Ulam-Collatz Sequence
95	[95, 286, 143, 430, 215, 646, 323, 970, 485, 1456, 728, 364, 182, 91, 274, 137, 412, 206, 103, 310, 155, 466, 233, 700, 350, 175, 526, 263, 790, 395, 1186, 593, 1780, 890, 445, 1336, 668, 334, 167, 502, 251, 754, 377, 1132, 566, 283, 850, 425, 1276, 638, 319, 958, 479, 1438, 719, 2158, 1079, 3238, 1619, 4858, 2429, 7288, 3644, 1822, 911, 2734, 1367, 4102, 2051, 6154, 3077, 9232, 4616, 2308, 1154, 577, 1732, 866, 433, 1300, 650, 325, 976, 488, 244, 122, 61, 184, 92, 46, 23, 70, 35, 106, 53, 160, 80, 40, 20, 10, 5, 16, 8, **4, 2, 1** (, 4)]
96	[96, 48, 24, 12, 6, 3, 10, 5, 16, 8, **4, 2, 1** (, 4)]
97	[97, 292, 146, 73, 220, 110, 55, 166, 83, 250, 125, 376, 188, 94, 47, 142, 71, 214, 107, 322, 161, 484, 242, 121, 364, 182, 91, 274, 137, 412, 206, 103, 310, 155, 466, 233, 700, 350, 175, 526, 263, 790, 395, 1186, 593, 1780, 890, 445, 1336, 668, 334, 167, 502, 251, 754, 377, 1132, 566, 283, 850, 425, 1276, 638, 319, 958, 479, 1438, 719, 2158, 1079, 3238, 1619, 4858, 2429, 7288, 3644, 1822, 911, 2734, 1367, 4102, 2051, 6154, 3077, 9232, 4616, 2308, 1154, 577, 1732, 866, 433, 1300, 650, 325, 976, 488, 244, 122, 61, 184, 92, 46, 23, 70, 35, 106, 53, 160, 80, 40, 20, 10, 5, 16, 8, **4, 2, 1** (, 4)]
98	[98, 49, 148, 74, 37, 112, 56, 28, 14, 7, 22, 11, 34, 17, 52, 26, 13, 40, 20, 10, 5, 16, 8, **4, 2, 1** (, 4)]
99	[99, 298, 149, 448, 224, 112, 56, 28, 14, 7, 22, 11, 34, 17, 52, 26, 13, 40, 20, 10, 5, 16, 8, **4, 2, 1** (, 4)]
100	[100, 50, 25, 76, 38, 19, 58, 29, 88, 44, 22, 11, 34, 17, 52, 26, 13, 40, 20, 10, 5, 16, 8, **4, 2, 1** (, 4)]

To get a graphic picture and see how some numbers appear to behave with this algorithm, we offer you the following examples.

Consider $n = 17$, which yields the loop: [17, 52, 26, 13, 40, 20, 10, 5, 16, 8, **4, 2, 1** (, 4)]. This requires 13 steps to reach the 1.

By comparison, $n = 18$ requires 20 steps to reach the 1.

Yet for $n = 27$, we would require 112 steps to reach 1. Figure 3-11 is a graphic representation of this progression.

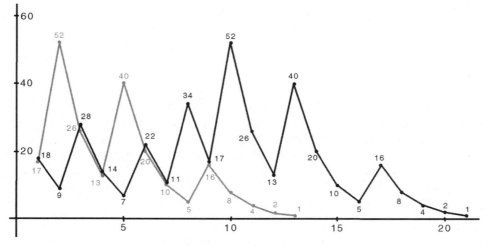

Figure 3-11 $n = 17$ and $n = 18$

For $n = 27$, the 1 is reached only after the following 112 steps:

[27, 82, 41, 124, 62, 31, 94, 47, 142, 71, 214, 107, 322, 161, 484, 242, 121, 364, 182, 91, 274, 137, 412, 206, 103, 310, 155, 466, 233, 700, 350, 175, 526, 263, 790, 395, 1186, 593, 1780, 890, 445, 1336, 668, 334, 167, 502, 251, 754, 377, 1132, 566, 283, 850, 425, 1276, 638, 319, 958, 479, 1438, 719, 2158, 1079, 3238, 1619, 4858, 2429, 7288, 3644, 1822, 911, 2734, 1367, 4102, 2051, 6154, 3077, 9232, 4616, 2308, 1154, 577, 1732, 866, 433, 1300, 650, 325, 976, 488, 244, 122, 61, 184, 92, 46, 23, 70, 35, 106, 53, 160, 80, 40, 20, 10, 5, 16, 8, **4, 2, 1** (, 4)]

Graphically, this would be depicted as shown in figure 3-12:

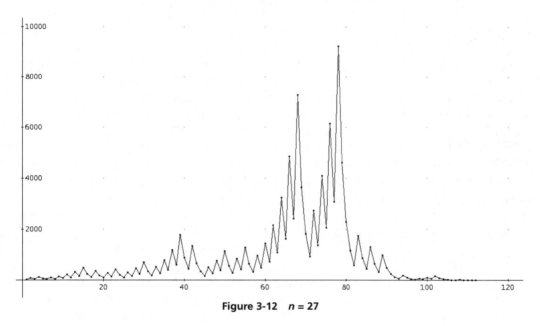

Figure 3-12   *n* = 27

For $n = 15{,}733{,}191$, we would require 705 steps to reach the 1, as shown here:

[15733191, 47199574, 23599787, 70799362, 35399681, 106199044, 53099522, 26549761, 79649284, 39824642, 19912321, 59736964, 29868482, 14934241, 44802724, 22401362, 11200681, 33602044, 16801022, 8400511, 25201534, 12600767, 37802302, 18901151, 56703454, 28351727, 85055182, 42527591, 127582774, 63791387, 191374162, 95687081, 287061244, 143530622, 71765311, 215295934, 107647967, 322943902, 161471951, 484415854, 242207927, 726623782, 363311891, 1089935674, 544967837, 1634903512, 817451756, 408725878, 204362939, 613088818, 306544409, 919633228, 459816614, 229908307, 689724922, 344862461, 1034587384, 517293692, 258646846, 129323423, 387970270, 193985135, 581955406, 290977703, 872933110, 436466555, 1309399666, 654699833, 1964099500, 982049750, 491024875, 1473074626, 736537313, 2209611940, 1104805970, 552402985,

1657208956, 828604478, 414302239, 1242906718, 621453359, 1864360078, 932180039, 2796540118, 1398270059, 4194810178, 2097405089, 6292215268, 3146107634, 1573053817, 4719161452, 2359580726, 1179790363, 3539371090, 1769685545, 5309056636, 2654528318, 1327264159, 3981792478, 1990896239, 5972688718, 2986344359, 8959033078, 4479516539, 13438549618, 6719274809, 20157824428, 10078912214, 5039456107, 15118368322, 7559184161, 22677552484, 11338776242, 5669388121, 17008164364, 8504082182, 4252041091, 12756123274, 6378061637, 19134184912, 9567092456, 4783546228, 2391773114, 1195886557, 3587659672, 1793829836, 896914918, 448457459, 1345372378, 672686189, 2018058568, 1009029284, 504514642, 252257321, 756771964, 378385982, 189192991, 567578974, 283789487, 851368462, 425684231, 1277052694, 638526347, 1915579042, 957789521, 2873368564, 1436684282, 718342141, 2155026424, 1077513212, 538756606, 269378303, 808134910, 404067455, 1212202366, 606101183, 1818303550, 909151775, 2727455326, 1363727663, 4091182990, 2045591495, 6136774486, 3068387243, 9205161730, 4602580865, 13807742596, 6903871298, 3451935649, 10355806948, 5177903474, 2588951737, 7766855212, 3883427606, 1941713803, 5825141410, 2912570705, 8737712116, 4368856058, 2184428029, 6553284088, 3276642044, 1638321022, 819160511, 2457481534, 1228740767, 3686222302, 1843111151, 5529333454, 2764666727, 8294000182, 4147000091, 12441000274, 6220500137, 18661500412, 9330750206, 4665375103, 13996125310, 6998062655, 20994187966, 10497093983, 31491281950, 15745640975, 47236922926, 23618461463, 70855384390, 35427692195, 106283076586, 53141538293, 159424614880, 79712307440, 39856153720, 19928076860, 9964038430, 4982019215, 14946057646, 7473028823, 22419086470, 11209543235, 33628629706, 16814314853, 50442944560, 25221472280, 12610736140, 6305368070, 3152684035, 9458052106, 4729026053, 14187078160, 7093539080, 3546769540, 1773384770, 886692385, 2660077156, 1330038578, 665019289, 1995057868, 997528934, 498764467, 1496293402, 748146701, 2244440104, 1122220052, 561110026, 280555013, 841665040, 420832520, 210416260, 105208130, 52604065, 157812196, 78906098, 39453049, 118359148, 59179574, 29589787, 88769362, 44384681, 133154044, 66577022, 33288511, 99865534, 49932767, 149798302, 74899151, 224697454, 112348727, 337046182, 168523091, 505569274, 252784637, 758353912, 379176956, 189588478, 94794239, 284382718, 142191359, 426574078, 213287039, 639861118, 319930559, 959791678, 479895839, 1439687518, 719843759, 2159531278, 1079765639, 3239296918, 1619648459, 4858945378, 2429472689, 7288418068, 3644209034, 1822104517, 5466313552, 2733156776, 1366578388, 683289194, 341644597, 1024933792, 512466896, 256233448, 128116724, 64058362, 32029181, 96087544, 48043772, 24021886, 12010943, 36032830, 18016415, 54049246, 27024623, 81073870, 40536935, 121610806, 60805403, 182416210, 91208105, 273624316, 136812158, 68406079, 205218238, 102609119, 307827358, 153913679, 461741038, 230870519, 692611558, 346305779, 1038917338, 519458669, 1558376008, 779188004, 389594002, 194797001, 584391004, 292195502, 146097751, 438293254, 219146627, 657439882, 328719941, 986159824, 493079912, 246539956, 123269978, 61634989, 184904968, 92452484, 46226242, 23113121, 69339364, 34669682, 17334841, 52004524, 26002262, 13001131, 39003394, 19501697, 58505092, 29252546, 14626273, 43878820, 21939410, 10969705, 32909116, 16454558, 8227279, 24681838, 12340919, 37022758, 18511379, 55534138, 27767069, 83301208, 41650604, 20825302, 10412651, 31237954, 15618977, 46856932, 23428466, 11714233, 35142700, 17571350, 8785675, 26357026, 13178513, 39535540, 19767770, 9883885,

29651656, 14825828, 7412914, 3706457, 11119372, 5559686, 2779843, 8339530, 4169765, 12509296, 6254648, 3127324, 1563662, 781831, 2345494, 1172747, 3518242, 1759121, 5277364, 2638682, 1319341, 3958024, 1979012, 989506, 494753, 1484260, 742130, 371065, 1113196, 556598, 278299, 834898, 417449, 1252348, 626174, 313087, 939262, 469631, 1408894, 704447, 2113342, 1056671, 3170014, 1585007, 4755022, 2377511, 7132534, 3566267, 10698802, 5349401, 16048204, 8024102, 4012051, 12036154, 6018077, 18054232, 9027116, 4513558, 2256779, 6770338, 3385169, 10155508, 5077754, 2538877, 7616632, 3808316, 1904158, 952079, 2856238, 1428119, 4284358, 2142179, 6426538, 3213269, 9639808, 4819904, 2409952, 1204976, 602488, 301244, 150622, 75311, 225934, 112967, 338902, 169451, 508354, 254177, 762532, 381266, 190633, 571900, 285950, 142975, 428926, 214463, 643390, 321695, 965086, 482543, 1447630, 723815, 2171446, 1085723, 3257170, 1628585, 4885756, 2442878, 1221439, 3664318, 1832159, 5496478, 2748239, 8244718, 4122359, 12367078, 6183539, 18550618, 9275309, 27825928, 13912964, 6956482, 3478241, 10434724, 5217362, 2608681, 7826044, 3913022, 1956511, 5869534, 2934767, 8804302, 4402151, 13206454, 6603227, 19809682, 9904841, 29714524, 14857262, 7428631, 22285894, 11142947, 33428842, 16714421, 50143264, 25071632, 12535816, 6267908, 3133954, 1566977, 4700932, 2350466, 1175233, 3525700, 1762850, 881425, 2644276, 1322138, 661069, 1983208, 991604, 495802, 247901, 743704, 371852, 185926, 92963, 278890, 139445, 418336, 209168, 104584, 52292, 26146, 13073, 39220, 19610, 9805, 29416, 14708, 7354, 3677, 11032, 5516, 2758, 1379, 4138, 2069, 6208, 3104, 1552, 776, 388, 194, 97, 292, 146, 73, 220, 110, 55, 166, 83, 250, 125, 376, 188, 94, 47, 142, 71, 214, 107, 322, 161, 484, 242, 121, 364, 182, 91, 274, 137, 412, 206, 103, 310, 155, 466, 233, 700, 350, 175, 526, 263, 790, 395, 1186, 593, 1780, 890, 445, 1336, 668, 334, 167, 502, 251, 754, 377, 1132, 566, 283, 850, 425, 1276, 638, 319, 958, 479, 1438, 719, 2158, 1079, 3238, 1619, 4858, 2429, 7288, 3644, 1822, 911, 2734, 1367, 4102, 2051, 6154, 3077, 9232, 4616, 2308, 1154, 577, 1732, 866, 433, 1300, 650, 325, 976, 488, 244, 122, 61, 184, 92, 46, 23, 70, 35, 106, 53, 160, 80, 40, 20, 10, 5, 16, 8, **4, 2, 1** (, 4)]

Graphically, this can be shown as in figure 3-13:

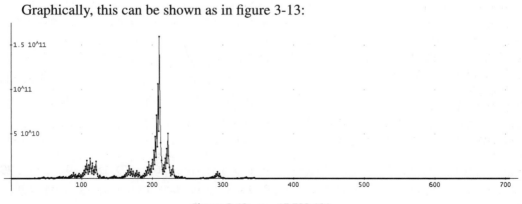

Figure 3-13 *n* = 15,733,191

The largest number reached in this process was $159,424,614,880 = 1.5942461488 \cdot 10^{11}$.

If you wish to try to perform this process for larger numbers, there is an Internet site that will do this calculation for you: http://math.carleton.ca/~amingare/mathzone/3*n*+1.html.

We don't want to discourage your further inspection of this curiosity. But we want to warn you not to get frustrated if you cannot prove that it is true in all cases, since we remind you that the best mathematical minds have not been able to do this after many attempts for the better part of a century!

A CYCLIC NUMBER LOOP

Take any integer from 1 to 6 and multiply it by 999,999 and then divide it by 7. You will get a number made up of the digits 1, 4, 2, 8, 5, 7. Not only that, but the digits will be in this order, yet starting from a different digit each time. This is the phenomenon of a *cyclic number*, which is a number of n digits that, when multiplied by each of the numbers 1, 2, 3, 4, ... , n, produces a number that uses the same digits as the original number but in a different order each time.[8]

Here is the result of multiplying the cyclic number 142,857 by each of the numbers 1 to 6 (see figure 3-14):

$142,857 \cdot 1 = 142,857$
$142,857 \cdot 2 = 285,714$
$142,857 \cdot 3 = 428,571$
$142,857 \cdot 4 = 571,428$
$142,857 \cdot 5 = 714,285$
$142,857 \cdot 6 = 857,142$

$142,857 \cdot 7 = \mathbf{999,999}$

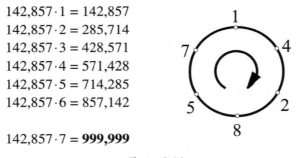

Figure 3-14

Notice that the pattern is also visible vertically through each decimal place.

Cyclic numbers are generated by certain prime numbers,[9] which, when they are the denominator of a unit fraction, yield a repetition of m digits, where m is one less than the prime. These types of prime numbers are called *full-reptend primes*.

The first few such full-reptend prime numbers are:

> 7, 17, 19, 23, 29, 47, 59, 61, 97, 109, 113, 131, 149, 167, 179, 181, 193, 223, 229, 233, 257, 263, 269, 313, 337, 367, 379, 383, 389, 419, 433, 461, 487, 491, 499, 503, 509, 541, 571, 577, 593, 619, 647, 659, 701, 709, 727, 743, 811, 821, 823, 857, 863, 887, 937, 941, 953, 971, 977, 983, . . .

---

8. These numbers are called *Phoenix numbers*—after the bird that according to an ancient Egyptian legend rises from its ashes whenever it is burned and disappears.

9. A *prime number* is an integer ($\neq 1$) that is divisible only by itself and 1.

A cyclic number is an $(n-1)$-digit integer that, when multiplied by 1, 2, 3, ... , $n-1$, produces the same digits in a different order. Therefore, cyclic numbers are generated by the full-reptend primes. To date, we still do not know if there are an infinite number of cyclic numbers.

Let's inspect the cyclic numbers generated by the first few of these primes. (The bar over the digits indicates that these digits repeat continuously, as, for example, we would write 6.18181818... as $6.\overline{18}$.)

$$\frac{1}{7} = 0.\overline{142857}$$

$$\frac{1}{17} = 0.\overline{0588235294117647}$$

$$\frac{1}{19} = 0.\overline{052631578947368421}$$

$$\frac{1}{23} = 0.\overline{0434782608695652173913}$$

$$\frac{1}{29} = 0.\overline{0344827586206896551724137931}$$

$$\frac{1}{47} = 0.\overline{0212765957446808510638297872340425531914893617}$$

$$\frac{1}{59} = 0.\overline{0169491525423728813559322033898305084745762711864406779661}$$

$$\frac{1}{61} = 0.\overline{016393442622950819672131147540983606557377049180327868852459}$$

It is this sequence of repeating digits that gives us the first few cyclic numbers. With the exception of the first cyclical number, we leave the zero at the beginning of the number for the first multiplication—it will not change the value of the number.

142857	(6 digits)
0588235294117647	(16 digits)
052631578947368421	(18 digits)
0434782608695652173913	(22 digits)
0344827586206896551724137931	(28 digits)
0212765957446808510638297872340425531914893617	(46 digits)
0169491525423728813559322033898305084745762711864406779661	(58 digits)
016393442622950819672131147540983606557377049180327868852459	(60 digits)

142857	· 7 = 999,999
0588235294117647	· 17 = 9,999,999,999,999,999
052631578947368421	· 19 = 999,999,999,999,999,999
0434782608695652173913	· 23 = 9,999,999,999,999,999,999,999
0344827586206896551724137931	· 29 = 9,999,999,999,999,999,999,999,999,999
0212765957446808510638297872340425531914893617	· 47 = 9,999,999,999,999,999,999,999,999,999,999,999,999,999,999,999
0169491525423728813559322033898305084745762711864406779661	· 59 = 9,999,999,999,999,999,999,999,999,999,999,999,999,999,999,999,999,999,999,999
016393442622950819672131147540983606557377049180327868852459	· 61 = 999,999,999,999,999,999,999,999,999,999,999,999,999,999,999,999,999,999,999,999

There is more that can be done with the cyclic number 142,857. Let's multiply it by some number, say, 1,818.

$$142,857 \cdot 1,818 = 259,714,026$$

If we take the right-most six digits—in this case 714,026—and add to that number the remaining digits—here 259—we get 714,285, which you will recognize as the cyclic number we began with. Here we have a true and recognizable loop.

You might also find another peculiarity of cyclic numbers. Consider again the cyclic number 142,857. We split it into two three-digit numbers and reverse the two halves and add them as:

$$142,857 + 857,142 = 999,999$$

Similarly, for the next cyclic number: 05882352,94117647 + 94117647,05882352 = 9999999999999999

We could also take the split numbers (as before) and add the two halves:

$$142 + 857 = 999$$
$$05882352 + 94117647 = 99999999$$
$$052631578 + 947368421 = 999999999$$

To qualify as a cyclic number, it is required that successive multiples be cyclic permutations. Thus, the number 076923 would not be considered a cyclic number, even though all cyclic permutations are multiples:

$$076923 \cdot 1 = 076923$$
$$076923 \cdot 3 = 230769$$
$$076923 \cdot 4 = 307692$$
$$076923 \cdot 9 = 692307$$
$$076923 \cdot 10 = 769230$$
$$076923 \cdot 12 = 923076$$

(Yet, for example, 076923·2 = 153846.)

A cyclic number may be created by taking the generating prime number and adding its doubles as shown below:

```
    1 4
        2 8
            5 6
          1 1 2
              2 2 4
                  4 4 8
                      8 9 6
                    1 7 9 2
                        3 5 8 4
                            7 1 6 8
                          1 4 3 3 6
                              2 8 6 7 2
    +                           5 7 3 4 4
                                        1
```
_____

1 4 2 8 5 7 1 4 2 8 5 7 **1 4 2 8 5 7** 1 4 2 8 ... ...

The number 865281023607, which is a multiple of 111111 (that is, 865281023607 = 111111·7787537), and all of the cyclic rotations of its digits are also multiples of 111111, as shown below:

$$865281023607 = 111111 \cdot 7787537$$
$$786528102360 = 111111 \cdot 7078760$$
$$078652810236 = 111111 \cdot 707876$$
$$607865281023 = 111111 \cdot 5470793$$
$$360786528102 = 111111 \cdot 3247082$$
$$236078652810 = 111111 \cdot 2124710$$
$$023607865281 = 111111 \cdot 212471$$
$$102360786528 = 111111 \cdot 921248$$
$$810236078652 = 111111 \cdot 7292132$$
$$281023607865 = 111111 \cdot 2529215$$
$$528102360786 = 111111 \cdot 4752926$$
$$652810236078 = 111111 \cdot 5875298$$

Furthermore, if we reverse the digits of these numbers, we find that each of them is again a multiple of 111111, as shown below:

$$706320182568 = 111111 \cdot 6356888$$
$$870632018256 = 111111 \cdot 7835696$$
$$687063201825 = 111111 \cdot 6183575$$
$$568706320182 = 111111 \cdot 5118362$$
$$256870632018 = 111111 \cdot 2311838$$
$$825687063201 = 111111 \cdot 7431191$$
$$182568706320 = 111111 \cdot 1643120$$
$$018256870632 = 111111 \cdot 164312$$
$$201825687063 = 111111 \cdot 1816433$$
$$320182568706 = 111111 \cdot 2881646$$
$$632018256870 = 111111 \cdot 5688170$$
$$063201825687 = 111111 \cdot 568817$$

CREATING A HALF-LOOP

There is a particular number that will exhibit many peculiarities, one of which is to bring everyone to the same place—not exactly a loop; so we will call it a "half-loop." This number is 1089. Let's first look at the first nine multiples of 1089.

$$1089 \cdot 1 = 1089$$
$$1089 \cdot 2 = 2178$$
$$1089 \cdot 3 = 3267$$
$$1089 \cdot 4 = 4356$$
$$1089 \cdot 5 = 5445$$
$$1089 \cdot 6 = 6534$$
$$1089 \cdot 7 = 7623$$
$$1089 \cdot 8 = 8712$$
$$1089 \cdot 9 = 9801$$

Do you notice a pattern among the products? Look at the first and ninth products (i.e., 1089 and 9801). They are the reverse of one another. The second and the eighth products (i.e., 2178 and 8712) are also the reverse of one another. And so, the pattern continues, until the fifth product, 5445, is the reverse of itself, known as a palindromic number.[10]

---

10. We had more about palindromic numbers on page 16.

A similar pattern occurs when we insert a 9 in the middle of the number 1089 to get 10,989. Notice how the reverses and the palindromes result when this number is multiplied by the first nine natural numbers.

$$10989 \cdot 1 = 10989$$
$$10989 \cdot 2 = 21978$$
$$10989 \cdot 3 = 32,967$$
$$10989 \cdot 4 = 43956$$
$$10989 \cdot 5 = 54945$$
$$10989 \cdot 6 = 65934$$
$$10989 \cdot 7 = 76923$$
$$10989 \cdot 8 = 87912$$
$$10989 \cdot 9 = 98901$$

Just as this unusual property holds true for 10989, it also holds true for 109989. You should recognize that we altered the original number, 1089, by inserting a 9 in the middle of the number to get 10989, and extended that by inserting 99 in the middle of the number 1089 to get 109989.

The following is the list of multiples of 10**99**89:

109989, 219978, 329967, 439956, 549945, 659934, 769923, 879912, 989901

It would be nice to conclude from this that each of the following numbers has the same property: 10**99**989, 10**999**989, 10**9999**989, 10**99999**989, 10**999999**989, and so on.

You will find that your calculation will verify this conjecture.

For the number 10**999**89, we have the following multiples:

1099989, 2199978, 3299967, 4399956, 5499945, 6599934, 7699923, 8799912, 9899901

You can try it with a calculator to find that this is true for all such numbers.

As a matter of fact, there is only one other number with four or fewer digits where a multiple of itself is equal to its reverse, and that is the number 2178, since $2178 \cdot 4 = 8712$. The number 2178 just happens to be $2 \cdot 1089$. Wouldn't it be nice if we could extend this, as we did with the above example, by inserting 9s into the middle of the number to generate other numbers that have the same property. Yes, it is true that

$$21978 \cdot 4 = 87912$$
$$219978 \cdot 4 = 879912$$
$$2199978 \cdot 4 = 8799912$$
$$21999978 \cdot 4 = 87999912$$
$$219999978 \cdot 4 = 879999912$$
$$2199999978 \cdot 4 = 8799999912$$
$$21999999978 \cdot 4 = 87999999912$$
$$219999999978 \cdot 4 = 879999999912$$
$$2199999999978 \cdot 4 = 8799999999912$$
$$21999999999978 \cdot 4 = 87999999999912$$
$$219999999999978 \cdot 4 = 879999999999912$$
$$2199999999999978 \cdot 4 = 8799999999999912$$
$$21999999999999978 \cdot 4 = 87999999999999912$$
$$219999999999999978 \cdot 4 = 879999999999999912$$

and so on.

THE 1089 LOOP

Now we go back to the original number, 1089. This number also provides us with a recreational phenomenon. We shall begin by having *you* select any three-digit number where the units digit and the hundreds digit are not the same.

Follow these instructions step by step, while we provide an example—below each of the instructions.

Choose any three-digit number
(where the units digit and the hundreds digit are not the same).

We will do it with you here by arbitrarily selecting **825**

Reverse the digits of this number you have selected.

We will continue here by reversing the digits of 825 to get **528**

Subtract the two numbers (naturally, the larger minus the smaller).

Our calculated difference is 825 − 528 = **297**

Once again, reverse the digits of this difference.

Reversing the digits of 297 we get the number **792**

Now, add your last two numbers.

We then add the last two numbers to get 297 + 792 = **1089**

Your result should be the same as ours (1089), even though your starting number was different from ours. (If not, then you made a calculation error. Check it.)

You will probably be astonished that regardless of which number you selected at the beginning, you got the same result as we did, 1089.

If the original three-digit number had the same units digit and hundreds digit, then we would get a zero after the first subtraction, such as for $n = 373$: $373 - 373 = 0$; this would ruin our model. Before reading on, convince yourself that this scheme will work for other numbers.

How does it happen? Is this a "freak property" of this number? Did we do something devious in our calculations?

This illustration of a mathematical oddity depends on the operations. We assumed that any number we chose would lead us to 1089. How can we be sure? Well, we could try all possible three-digit numbers to see if it works. That would be tedious and not particularly elegant. An investigation of this oddity requires nothing more than some knowledge of elementary algebra.

Yet were we to try to test all possibilities, it would be interesting to determine to how many such three-digit numbers we would have to apply this scheme. Remember, we can only use those three-digit numbers whose units digit and hundreds digit are not the same.

If we consider all the three-digit numbers from 100 to 199, we find—by eliminating those that have equal units and hundreds digits—that there are ninety such numbers.

100	110	120	130	140	150	160	170	180	190
~~101~~	~~111~~	~~121~~	~~131~~	~~141~~	~~151~~	~~161~~	~~171~~	~~181~~	~~191~~
102	112	122	132	142	152	162	172	182	192
103	113	123	133	143	153	163	173	183	193
104	114	124	134	144	154	164	174	184	194
105	115	125	135	145	155	165	175	185	195
106	116	126	136	146	156	166	176	186	196
107	117	127	137	147	157	167	177	187	197
108	118	128	138	148	158	168	178	188	198
109	119	129	139	149	159	169	179	189	199

The same is true for the three-digit numbers from 200 to 299, and from 300 to 399, and from 400 to 499, and so on. The total of such numbers is $9 \cdot 90 = 810$, which would have to be tested.

For the reader who might be curious about this phenomenon and doesn't want to go through this tedious testing process, we will provide an (elementary) algebraic explanation as to why it "works." We simply apply our understanding of the decimal place-value system.

We shall represent the arbitrarily selected three-digit number, $n = \overline{htu}$, as $h \cdot 100 + t \cdot 10 + u$, where h represents the hundreds digit, t represents the tens digit, and u represents the units digit, for example, $\mathbf{825 = 8 \cdot 100 + 2 \cdot 10 + 5 \cdot 1 = 8 \cdot 10^2 + 2 \cdot 10^1 + 5 \cdot 10^0}$.

In this case, $h = 8$, $t = 2$, $u = 5$, and then the number is represented as $n = \overline{htu}$ and the reversal of n can be represented as $r(n) = \overline{uth}$.

Let $h > u$, which would be the case either in the number n you selected or the reverse $r(n)$ of it.

In the subtraction, $u - h < 0$ ($5 - 8 < 0$); therefore, take 1 from the tens place (of the minuend) making the units place $10 + u$.

In our illustration this appears as:

$$
\begin{array}{cccc}
825 & 800 + 20 + 5 & 8\;1\;15 & 800 + 10 + 15 \\
-528 & & -5\;2\;8 & \\
\end{array}
$$

$$
\begin{array}{cccc}
7\;\;11\;\;15 & 700 + 110 + 15 & 7\;\;11\;\;15 & \\
-5\;\;2\;\;8 & & -5\;\;2\;\;8 & \\
& & 2\;9\;7 & \\
\end{array}
$$

Since the tens digits of the two numbers to be subtracted are equal, and 1 was taken from the tens digit of the minuend n, then the value of this digit is $(t - 1) \cdot 10$. The hundreds digit of the minuend n is $h - 1$, because 1 was taken away to enable subtraction in the tens place, making the value of the tens digit $(10 + t - 1) \cdot 10 = (t + 9) \cdot 10$.

We can now do the first subtraction:

$$
\begin{array}{ccccc}
 & (h - 1) \cdot 100 & + & (t + 9) \cdot 10 & + & u + 10 \\
- & u \cdot 100 & + & t \cdot 10 & + & h \\
\hline
 & (h - u - 1) \cdot 100 & + & 9 \cdot 10 & + & u + 10 - h \\
\end{array}
$$

Reversing the digits of this difference $(h - u - 1) \cdot 100 + 9 \cdot 10 + (u + 10 - h)$ gives us: $(u + 10 - h) \cdot 100 + 9 \cdot 10 + (h - u - 1)$.

Now adding these last two expressions gives us:

$$(h - u - 1) \cdot 100 + 9 \cdot 10 + (u + 10 - h) + (u + 10 - h) \cdot 100 + 9 \cdot 10 + (h - u - 1)$$
$$= \underline{100h} - \underline{100u} - 100 + 90 \underline{+u} + 10 \underline{-h} + \underline{100u} + 1{,}000 - \underline{100h} + 90 \underline{+h} - \underline{u} - 1$$
$$= \underline{-100 + 90 + 10} + 1{,}000 + 90 - 1 = 1{,}090 - 1 = \mathbf{1089}$$

The true value of this algebraic justification is that it enables us to inspect the arithmetic process, regardless of the number selected.

Before we leave the number 1089, we should point out that there is still another neat feature of this number. A relationship we recognize first as a Pythagorean one: $33^2 + 56^2 = 65^2$, when written in this form $33^2 = 1089 = 65^2 - 56^2$, has some lovely symmetry to it that is unique among two-digit numbers. By now, you must agree that there is a particular beauty in the number **1089**.

At this point, you may be wondering if this sort of scheme might also hold true for larger numbers, which is a reasonable question. If we begin with any four-digit number where the units digit and the thousands digit are not the same, we might get a constant number as we did before, but then again, we might not. Let's inspect this situation more closely.

We shall follow the earlier instructions for three-digit numbers, but this time we will use it for four-digit numbers:

Choose any four-digit number (this time without restrictions).
We will do it with you here by arbitrarily selecting **8,029**

Reverse the digits of this arbitrarily selected number.
We will continue here by reversing the digits of 8,029 to get **9,208**

Subtract the two numbers (naturally, the larger minus the smaller).
Our calculated difference is 9,208 – 8,029 = **1,179**

Once again, reverse the digits of this difference.
Reversing the digits of 1,179 we get the number **9,711**

Now, add your last two numbers.
We then add the last two numbers to get 1,179 + 9,711 = **10,890**

At first sight, this appears to be consistent with our earlier situation. As you will see from the list below, the mere restriction that the units digit and the thousands digit not be the same will not be enough to guarantee our reaching the number 10,890.

For each value of n, we may or may not reach the desired number 10,890. You can see from the chart below that there is a pattern of numbers reached by applying this procedure to numbers 8,100–8,200. This will give you some insight into the nature of the situation. You might investigate the results with other four-digit numbers to see if the pattern is consistent.

n	Procedure result	n	Procedure result
8,100	10,890	8,161	**9,999**
8,101	10,890	...	...
8,102	10,890	8,167	**9,999**
...	...	8,168	**990**
8,107	10,890	8,169	10,890
8,108	**990**	8,170	**9,999**
8,109	**9,999**	...	...
8,110	10,989	8,177	**9,999**
8,111	19,889	8,178	**990**
...	...	8,179	10,890
8,117	10,989		
8,118	**0**		
8,119	10,989	8,189	10,890
8,120	**9,999**	8,190	**9,999**
8,121	**9,999**	8,191	**9,999**
...	...	...	...
8,127	**9,999**	8,197	**9,999**
8,128	**990**	8,198	**990**
8,129	10,890	8,199	10,890
8,130	**9,999**	8,200	10,890
...	...		
8,138	**9,999**		
8,139	10,890		

This will further aggrandize the amazing result we found for the three-digit number application!

Let's see what happens if we try this procedure with two-digit numbers: Follow this:

Choose any two-digit number with different digits.
We will do it with you here by arbitrarily selecting **37**

Reverse the digits of this arbitrarily selected number.
We will continue here by reversing the digits of 37 to get **73**

Subtract the two numbers (naturally, the larger minus the smaller).
Our calculated difference is 73 − 37 = **36**

Once again, reverse the digits of this difference.
Reversing the digits of 36, we get the number **63**

Now, add your last two numbers.
We then add the last two numbers to get 36 + 63 = **99**

You ought to be able to justify a pattern just as we did for the previous four-digit application of the procedure.

THE 99 LOOP

Now we vary the earlier procedure that led to the number 1089 (which we called the "1089 half-loop").

That means that we shall again begin with any three-digit number where the units digit and the hundreds digit are not the same.

Choose any three-digit number
(where the units digit and the hundreds digit are not the same).
We will do it with you here by arbitrarily selecting **825**

Reverse the digits of this number you have selected.
We will continue here by reversing the digits of 825 to get **528**

Subtract the two numbers (naturally, the larger minus the smaller).
Our calculated difference is 825 − 528 = **297**

Once again, reverse the digits of this difference.
Reversing the digits of 297 we get the number **792**

Then continue this process of subtracting and reversing.

$$792 - 297 = \mathbf{495},$$
$$594 - 495 = 99,$$
$$990 - 099 = 891,$$
$$891 - 198 = 693,$$
$$693 - 396 = 297,$$
$$792 - 297 = \mathbf{495}, \ldots$$

In this variation of the original scheme, we do not reach a constant number (1089), rather we end up in a loop of length 5: [99, 891, 693, 297, 495 (, 99)].

In the case where the units digit and the hundreds digit are the same, we end up in a loop of length 1: [0].

Let's look at some examples. Using, for illustrative purposes, the numbers 771, 240, and 102:

771	240	102
− 177	− 042	
594	198	201
− **495**		− 102
099	**891**	**099**
	− 198	
990	**693**	990
− 099	− 396	− 099
891	**297**	**891**
− 198		− 198
693	792	**693**
− 396	− 297	− 396
297	**495**	**297**
792	594	792
− 297	− 495	− 297
495	099	**495**
	990	594
	− 099	− 495
	891	**099**

Suppose we consider a single-digit number, such as 3, to be a three-digit number written as 003. We then have the possibility of inspecting smaller than traditional three-digit numbers to see how this scheme works on these smaller numbers. So, for the number 7, which we would now write as 007, we begin the process by subtracting: $700 - 007 = 693$.

Then, continuing the process, we get: 693 – 396 = 297, which then leads us to 792 – 297 = 495, then 594 – 495 = 99, then 990 – 099 = 891, then 891 – 198 = 693, which completes the loop!

We can see this graphically with figure 3-15 below, where we show the loops for the numbers from 001 to 025.

Visualization of the sequences for $n = 1, 2, 3, \ldots, 25$
with the loop [99, 891, 693, 297, 495 (, 99)].

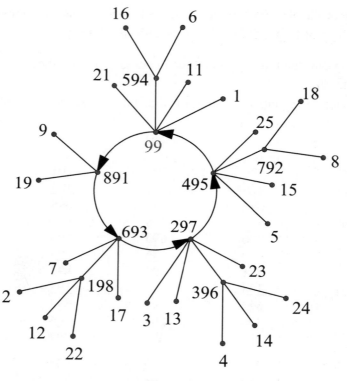

Figure 3-15

The sequences for the first one hundred numbers are shown in the chart on the following page so you can see how the loops develop.

Sequences of the process for *n* = 1, 2, ... , 100

n	Sequence	*n*	Sequence
1	[1, 99, 891, 693, 297, 495, 99]	51	[51, 99, 891, 693, 297, 495, 99]
2	[2, 198, 693, 297, 495, 99, 891, 693]	52	[52, 198, 693, 297, 495, 99, 891, 693]
3	[3, 297, 495, 99, 891, 693, 297]	53	[53, 297, 495, 99, 891, 693, 297]
4	[4, 396, 297, 495, 99, 891, 693, 297]	54	[54, 396, 297, 495, 99, 891, 693, 297]
5	[5, 495, 99, 891, 693, 297, 495]	55	[55, 495, 99, 891, 693, 297, 495]
6	[6, 594, 99, 891, 693, 297, 495, 99]	56	[56, 594, 99, 891, 693, 297, 495, 99]
7	[7, 693, 297, 495, 99, 891, 693]	57	[57, 693, 297, 495, 99, 891, 693]
8	[8, 792, 495, 99, 891, 693, 297, 495]	58	[58, 792, 495, 99, 891, 693, 297, 495]
9	[9, 891, 693, 297, 495, 99, 891]	59	[59, 891, 693, 297, 495, 99, 891]
10	[10, 0, 0]	60	[60, 0, 0]
11	[11, 99, 891, 693, 297, 495, 99]	61	[61, 99, 891, 693, 297, 495, 99]
12	[12, 198, 693, 297, 495, 99, 891, 693]	62	[62, 198, 693, 297, 495, 99, 891, 693]
13	[13, 297, 495, 99, 891, 693, 297]	63	[63, 297, 495, 99, 891, 693, 297]
14	[14, 396, 297, 495, 99, 891, 693, 297]	64	[64, 396, 297, 495, 99, 891, 693, 297]
15	[15, 495, 99, 891, 693, 297, 495]	65	[65, 495, 99, 891, 693, 297, 495]
16	[16, 594, 99, 891, 693, 297, 495, 99]	66	[66, 594, 99, 891, 693, 297, 495, 99]
17	[17, 693, 297, 495, 99, 891, 693]	67	[67, 693, 297, 495, 99, 891, 693]
18	[18, 792, 495, 99, 891, 693, 297, 495]	68	[68, 792, 495, 99, 891, 693, 297, 495]
19	[19, 891, 693, 297, 495, 99, 891]	69	[69, 891, 693, 297, 495, 99, 891]
20	[20, 0, 0]	70	[70, 0, 0]
21	[21, 99, 891, 693, 297, 495, 99]	71	[71, 99, 891, 693, 297, 495, 99]
22	[22, 198, 693, 297, 495, 99, 891, 693]	72	[72, 198, 693, 297, 495, 99, 891, 693]
23	[23, 297, 495, 99, 891, 693, 297]	73	[73, 297, 495, 99, 891, 693, 297]
24	[24, 396, 297, 495, 99, 891, 693, 297]	74	[74, 396, 297, 495, 99, 891, 693, 297]
25	[25, 495, 99, 891, 693, 297, 495]	75	[75, 495, 99, 891, 693, 297, 495]
26	[26, 594, 99, 891, 693, 297, 495, 99]	76	[76, 594, 99, 891, 693, 297, 495, 99]
27	[27, 693, 297, 495, 99, 891, 693]	77	[77, 693, 297, 495, 99, 891, 693]
28	[28, 792, 495, 99, 891, 693, 297, 495]	78	[78, 792, 495, 99, 891, 693, 297, 495]
29	[29, 891, 693, 297, 495, 99, 891]	79	[79, 891, 693, 297, 495, 99, 891]
30	[30, 0, 0]	80	[80, 0, 0]
31	[31, 99, 891, 693, 297, 495, 99]	81	[81, 99, 891, 693, 297, 495, 99]
32	[32, 198, 693, 297, 495, 99, 891, 693]	82	[82, 198, 693, 297, 495, 99, 891, 693]
33	[33, 297, 495, 99, 891, 693, 297]	83	[83, 297, 495, 99, 891, 693, 297]
34	[34, 396, 297, 495, 99, 891, 693, 297]	84	[84, 396, 297, 495, 99, 891, 693, 297]
35	[35, 495, 99, 891, 693, 297, 495]	85	[85, 495, 99, 891, 693, 297, 495]
36	[36, 594, 99, 891, 693, 297, 495, 99]	86	[86, 594, 99, 891, 693, 297, 495, 99]
37	[37, 693, 297, 495, 99, 891, 693]	87	[87, 693, 297, 495, 99, 891, 693]
38	[38, 792, 495, 99, 891, 693, 297, 495]	88	[88, 792, 495, 99, 891, 693, 297, 495]
39	[39, 891, 693, 297, 495, 99, 891]	89	[89, 891, 693, 297, 495, 99, 891]
40	[40, 0, 0]	90	[90, 0, 0]
41	[41, 99, 891, 693, 297, 495, 99]	91	[91, 99, 891, 693, 297, 495, 99]
42	[42, 198, 693, 297, 495, 99, 891, 693]	92	[92, 198, 693, 297, 495, 99, 891, 693]
43	[43, 297, 495, 99, 891, 693, 297]	93	[93, 297, 495, 99, 891, 693, 297]
44	[44, 396, 297, 495, 99, 891, 693, 297]	94	[94, 396, 297, 495, 99, 891, 693, 297]
45	[45, 495, 99, 891, 693, 297, 495]	95	[95, 495, 99, 891, 693, 297, 495]
46	[46, 594, 99, 891, 693, 297, 495, 99]	96	[96, 594, 99, 891, 693, 297, 495, 99]
47	[47, 693, 297, 495, 99, 891, 693]	97	[97, 693, 297, 495, 99, 891, 693]
48	[48, 792, 495, 99, 891, 693, 297, 495]	98	[98, 792, 495, 99, 891, 693, 297, 495]
49	[49, 891, 693, 297, 495, 99, 891]	99	[99, 891, 693, 297, 495, 99]
50	[50, 0, 0]	100	[100, 99, 891, 693, 297, 495, 99]

The situation changes dramatically when we consider four-digit numbers. To inspect this case, we shall again begin with any four-digit number without restriction (that also means that all digits could be the same).

Choose any four-digit number.

We will do it with you here by arbitrarily selecting **3,795**

Reverse the digits of this number you have selected.

We will continue here by reversing the digits of 3,795 to get **5,973**

Subtract the two numbers (naturally, the larger minus the smaller).

Our calculated difference is 5,973 – 3,795 = **2,178**

Once again, reverse the digits of this difference.

Reversing the digits of 2,178 we get the number **8,712**

Then continue this process of subtracting and reversing.

$$8,712 - 2,178 = 6,534,$$
$$6,534 - 4,356 = 2,178, \dots$$

Notice the loop!

Another Example: $n = 9,916$

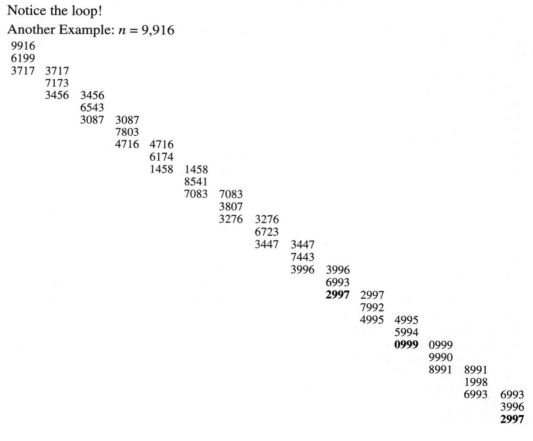

In short form we arrive at this sequence:

> [9916, 3717, 3456, 3087, 4716, 1458, 7083, 3276, 3447, 3996, 2997, 4995, **999**, 8991, 6993, 2997 (, 4995, 999)]

In contrast to the earlier situation with three-digit numbers, where we landed in a common loop, here—as you can see from the example—we land in a loop of length 5: [999, 8991, 6993, 2997, 4995]. However, we can actually land in five possible loops:

The loop [0] of length 1	· (For example, $n = 2{,}277$)
The loop [2178, 6534] of length 2	(For example, $n = 3{,}795$)
The loop [90, 810, 630, 270, 450] of length 5	(For example, $n = 3{,}798$)
The loop [909, 8181, 6363, 2727, 4545] of length 5	(For example, $n = 5{,}514$)
The loop [999, 8991, 6993, 2997, 4995] of length 5	(For example, $n = 382$)

By the way, the loop [0] does not only occur for repeating digits as above; the starting number 1,056 also ends in a 0: [1056, 5445, 0, 0].

We offer, here, the results of the process on the following values of n:
$(1{,}100 \leq n \leq 1{,}200)$

[1100, 1089, 8712, 6534, **2178**],
[1101, **90**],
[1102, **909**],
[1103, 1908, 6183, 2367, 5265, 360, 270, 450, **90**],
[1104, 2907, 4185, 1629, 7632, 5265, 360, 270, 450, **90**],
[1105, 3906, 2187, 5625, 360, 270, 450, **90**],
[1106, 4905, 189, 9621, 8352, 5814, 1629, 7632, 5265, 360, 270, 450, **90**],
[1107, 5904, 1809, 7272, 4545, **909**],
[1108, 6903, 3807, 3276, 3447, 3996, 2997, 4995, **999**],
[1109, 7902, 5805, 720, 450, **90**],
[1110, **999**],
[1111, **0**],
[1112, **999**],
[1113, 1998, 6993, 2997, 4995, **999**],
[1114, 2997, 4995, **999**],
[1115, 3996, 2997, 4995, **999**],
[1116, 4995, **999**],
[1117, 5994, **999**],
[1118, 6993, 2997, 4995, **999**],
[1119, 7992, 4995, **999**],
[1120, **909**],

[1121, **90**],

[1122, 1089, 8712, 6534, **2178**],

[1123, 2088, 6714, 2538, 5814, 1629, 7632, 5265, 360, 270, 450, **90**],

[1124, 3087, 4716, 1458, 7083, 3276, 3447, 3996, 2997, 4995, **999**],

[1125, 4086, 2718, 5454, **909**],

[1126, 5085, 720, 450, **90**],

[1127, 6084, 1278, 7443, 3996, 2997, 4995, **999**],

[1128, 7083, 3276, 3447, 3996, 2997, 4995, **999**],

[1129, 8082, 5274, 549, 8901, 7803, 4716, 1458, 7083, 3276, 3447, 3996, 2997, 4995, **999**],

[1130, 819, 8361, 6723, 3447, 3996, 2997, 4995, **999**],

[1131, 180, 630, 270, 450, **90**],

[1132, 1179, 8532, 6174, 1458, 7083, 3276, 3447, 3996, 2997, 4995, **999**],

[1133, **2178**],

[1134, 3177, 4536, 1818, 6363, 2727, 4545, **909**],

[1135, 4176, 2538, 5814, 1629, 7632, 5265, 360, 270, 450, **90**],

[1136, 5175, 540, **90**],

[1137, 6174, 1458, 7083, 3276, 3447, 3996, 2997, 4995, **999**],

[1138, 7173, 3456, 3087, 4716, 1458, 7083, 3276, 3447, 3996, 2997, 4995, **999**],

[1139, 8172, 5454, **909**],

[1140, 729, 8541, 7083, 3276, 3447, 3996, 2997, 4995, **999**],

[1141, 270, 450, **90**],

[1142, 1269, 8352, 5814, 1629, 7632, 5265, 360, 270, 450, **90**],

[1143, 2268, 6354, 1818, 6363, 2727, 4545, **909**],

[1144, 3267, 4356, **2178**],

[1145, 4266, 2358, 6174, 1458, 7083, 3276, 3447, 3996, 2997, 4995, **999**],

[1146, 5265, 360, 270, 450, **90**],

[1147, 6264, 1638, 6723, 3447, 3996, 2997, 4995, **999**],

[1148, 7263, 3636, 2727, 4545, **909**],

[1149, 8262, 5634, 1269, 8352, 5814, 1629, 7632, 5265, 360, 270, 450, **90**],

[1150, 639, 8721, 7443, 3996, 2997, 4995, **999**],

[1151, 360, 270, 450, **90**],

[1152, 1359, 8172, 5454, **909**],

[1153, 2358, 6174, 1458, 7083, 3276, 3447, 3996, 2997, 4995, **999**],

[1154, 3357, 4176, 2538, 5814, 1629, 7632, 5265, 360, 270, 450, **90**],

[1155, 4356, **2178**],

[1156, 5355, 180, 630, 270, 450, **90**],

[1157, 6354, 1818, 6363, 2727, 4545, **909**],

[1158, 7353, 3816, 2367, 5265, 360, 270, 450, **90**],

[1159, 8352, 5814, 1629, 7632, 5265, 360, 270, 450, **90**],

[1160, 549, 8901, 7803, 4716, 1458, 7083, 3276, 3447, 3996, 2997, 4995, **999**],

[1161, 450, **90**],

[1162, 1449, 7992, 4995, **999**],

[1163, 2448, 5994, **999**],

[1164, 3447, 3996, 2997, 4995, **999**],

[1165, 4446, 1998, 6993, 2997, 4995, **999**],
[1166, 5445, **0**],
[1167, 6444, 1998, 6993, 2997, 4995, **999**],
[1168, 7443, 3996, 2997, 4995, **999**],
[1169, 8442, 5994, **999**],
[1170, 459, 9081, 7272, 4545, **909**],
[1171, 540, **90**],
[1172, 1539, 7812, 5625, 360, 270, 450, **90**],
[1173, 2538, 5814, 1629, 7632, 5265, 360, 270, 450, **90**],
[1174, 3537, 3816, 2367, 5265, 360, 270, 450, **90**],
[1175, 4536, 1818, 6363, 2727, 4545, **909**],
[1176, 5535, 180, 630, 270, 450, **90**],
[1177, 6534, **2178**],
[1178, 7533, 4176, 2538, 5814, 1629, 7632, 5265, 360, 270, 450, **90**],
[1179, 8532, 6174, 1458, 7083, 3276, 3447, 3996, 2997, 4995, **999**],
[1180, 369, 9261, 7632, 5265, 360, 270, 450, **90**],
[1181, 630, 270, 450, **90**],
[1182, 1629, 7632, 5265, 360, 270, 450, **90**],
[1183, 2628, 5634, 1269, 8352, 5814, 1629, 7632, 5265, 360, 270, 450, **90**],
[1184, 3627, 3636, 2727, 4545, **909**],
[1185, 4626, 1638, 6723, 3447, 3996, 2997, 4995, **999**],
[1186, 5625, 360, 270, 450, **90**],
[1187, 6624, 2358, 6174, 1458, 7083, 3276, 3447, 3996, 2997, 4995, **999**],
[1188, 7623, 4356, **2178**],
[1189, 8622, 6354, 1818, 6363, 2727, 4545, **909**],
[1190, 279, 9441, 7992, 4995, **999**],
[1191, 720, 450, **90**],
[1192, 1719, 7452, 4905, 189, 9621, 8352, 5814, 1629, 7632, 5265, 360, 270, 450, **90**],
[1193, 2718, 5454, **909**],
[1194, 3717, 3456, 3087, 4716, 1458, 7083, 3276, 3447, 3996, 2997, 4995, **999**],
[1195, 4716, 1458, 7083, 3276, 3447, 3996, 2997, 4995, **999**],
[1196, 5715, 540, **90**],
[1197, 6714, 2538, 5814, 1629, 7632, 5265, 360, 270, 450, **90**],
[1198, 7713, 4536, 1818, 6363, 2727, 4545, **909**],
[1199, 8712, 6534, **2178**],
[1200, 1179, 8532, 6174, 1458, 7083, 3276, 3447, 3996, 2997, 4995, **999**]

A FACTORIAL LOOP

This charming little loop will show an unusual relationship for certain numbers. But before beginning, let us recall the definition of $n!$ as $n! = 1 \cdot 2 \cdot 3 \cdot 4 \cdot \ldots \cdot (n - 1) \cdot n$. For consistency we set $0! = 1$.

To check your understanding of the factorial concept, find the sum of the factorials of the digits of 145.

$$1! + 4! + 5! = 1 + 24 + 120 = \textbf{145}$$

Surprise! We're back to 145.

Only for certain numbers will the sum of the factorials of the digits equal the number itself.

Try this again with the number 40,585.

That is, $4! + 0! + 5! + 8! + 5! = 24 + 1 + 120 + 40,320 + 120 = \textbf{40,585}$.

At this point, you may expect this to be true for just about any number. Well, just try another number. Chances are that it may not work.

Now, try this scheme with the number 871. You will get: $8! + 7! + 1! = 40,320 + 5,040 + 1 = 45,361$, at which point you may feel that you have failed. Not so fast. Try this procedure again with 45,361. This will give you: $4! + 5! + 3! + 6! + 1! = 24 + 120 + 6 + 720 + 1 = 871$. Isn't this the very number we started with? Again, we formed a loop.

If you repeat this with the number 872, you will get $8! + 7! + 2! = 40,320 + 5,040 + 2 = 45,362$. Then, repeating the process with this number will give you: $4! + 5! + 3! + 6! + 2! = 24 + 120 + 6 + 720 + 2 = \textbf{872}$. Again, we are in a loop.

Some people are quick to form generalizations, so they might conclude that if the scheme of summing factorials of the digits of a number doesn't get them back to the original number, then they will try it again, thinking it ought to work. Of course they can "stack the deck" by using the number 169 as their next try. Two cycles do not seem to present a loop. So then if they proceed through one more cycle, sure enough, the third cycle will lead back to the original number.

Starting number	Sum of the factorials
169	$1! + 6! + 9! = 1 + 720 + 362,880 = 363,601$
363,601	$3! + 6! + 3! + 6! + 0! + 1! = 6 + 720 + 6 + 720 + 1 + 1 = 1,454$
1,454	$1! + 4! + 5! + 4! = 1 + 24 + 120 + 24 = \textbf{169}$

Be careful about drawing conclusions. These factorial oddities are not so pervasive that you should be able to easily find others. There are "within reach" three groups of such loops. We can organize them according to the number of times you have to repeat the process to reach the original number. We will call these repetitions "cycles."

Here is a summary of the way our numbers behave in this factorial loop.

1 cycle	[1], [2], [145], [40585]
2 cycle	[871, 45361] and [872, 45362]
3 cycle	[169, 363601, 1454]

Let's revisit the ever-popular Armstrong number 153.

$$1! + 5! + 3! = 127$$
$$1! + 2! + 7! = 5{,}043$$
$$5! + 0! + 4! + 3! = 151$$
$$1! + 5! + 1! = 122$$
$$1! + 2! + 2! = 5$$
$$5! = 120$$
$$1! + 2! + 0! = 4$$
$$4! = 24$$
$$2! + 4! = 26$$
$$2! + 6! = 722$$
$$7! + 2! + 2! = 5{,}044$$
$$5! + 0! + 4! + 4! = \mathbf{169}$$
$$1! + 6! + 9! = \mathbf{363{,}601}$$
$$3! + 6! + 3! + 6! + 0! + 1! = \mathbf{1{,}454}$$
$$1! + 4! + 5! + 4! = \mathbf{169}$$

We arrive at the loop [169, 363601, 1454] of length 3.

The factorial loops shown in this charming little number oddity can be fun, but be aware that there are no other such numbers less than 2,000,000 for which this works. So just appreciate these little beauties for what they are!

We have now witnessed an astounding array of number peculiarities that should give you motivation to seek others. Mathematics holds a treasure trove of such number patterns that can be discovered through a clear imagination and some perseverance.

Chapter 4

GEOMETRY SURPRISES

Then the topic of geometry evokes, in many people, their high school memories of the subject. Usually, these recollections encompass the many proofs they were required to do to establish geometric relationships. Too often, the process of the proofs became the focus, overshadowing the actual relationships being proved and, unfortunately, leaving the beauty of geometry somewhat neglected. This chapter is intended to reverse that process. In order to demonstrate the beauty of geometry, we will present some truly awe-inspiring relationships—all of which (obviously) can be proved—but, so as not to disturb the "flow" of the presentation, we will merely refer you to sources where the proofs can be found. (See the bibliography for some suggested books.)

Join us now as we explore some truly amazing geometric relationships. However, before we embark on our journey through these geometric wonders, we must caution you that appearance alone can be misleading. Sometimes what appears to be true is not, and what appears impossible can be possible. This makes our journey through geometry that much more intriguing. To demonstrate that, we offer some optical illusions for your entertainment.

OPTICAL ILLUSIONS

In geometry, what you see is not always true. For example, in figure 4-1 you see two white squares. Most observers would say that the left-side white square is a bit smaller than the white square on the right. This is incorrect. They are both the same size. Their context and presentation make for an optical illusion.

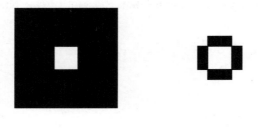

Figure 4-1

Another example to show you that your optical judgment may not be as accurate as you might think is that of figure 4-2, where the circle inscribed in the square on the left appears to be smaller than the circle circumscribed about the square on the right. Again, this is not true, since they are the same size.

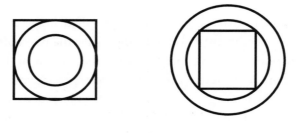

Figure 4-2

To further "upset" your sense of perception, consider the following few optical illusions. In figure 4-3, the center (black) circle on the left appears to be smaller than the center (black) circle on the right. Again, not so. They are both of equal size.

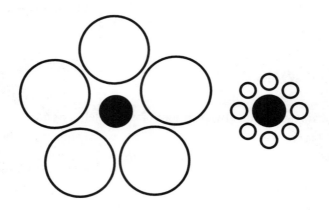

Figure 4-3

In figure 4-4, *AB* appears to be longer than *BC*, but it isn't. They are the same length.

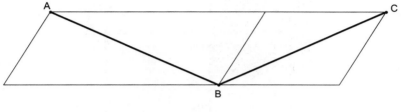

Figure 4-4

In figure 4-5, the horizontal line appears to be shorter than the vertical line, but, again, you have an optical illusion, since they are actually the same length.

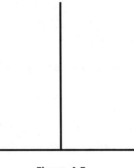

Figure 4-5

Some optical illusions are done intentionally. Take, for example, the *tribar* or *Penrose triangle* (figure 4-6). Popularized by Roger Penrose (1931–) in 1958, it was previously created in 1934 by Oscar Reutersvärd (1915–2002), a Swedish artist. It appears to be a triangle with three right angles. His invention was honored by the issuance of a postage stamp in 1982 (see figure 4-7).[1]

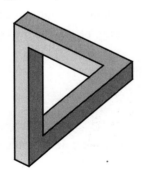

Figure 4-6

Figure 4-7

---

1. Swedish postage stamp: 25 Öre, Sverige, February 16, 1982.

A postage stamp depicting an "irrational" cube was issued by the Republic of Austria in 1981 to commemorate the tenth International Mathematics Congress in Innsbruck (see figure 4-8).

Figure 4-8

Just as there are optical illusions throughout our geometric world, so too there are "proofs" that can be fallacious, not by their errors in reasoning, but rather in their assumptions of geometric appearances.

A very popular "fallacy proof" is that every triangle can be proved to be isosceles.[2] The error in the proof is one that might have baffled Euclid, who did not properly define the concept of betweenness.[3]

Do not let these unusual optical illusions and tricks in geometry disillusion you. There is much beauty in geometry that is true. By this we are not referring to visually beautiful things, rather we are referring to amazing relationships that you would not think possible. We will present some here to rekindle your curiosity and to enhance your perception of geometry as more than just something that is optically interesting or just a series of proofs.

LOOKING AT GEOMETRY CRITICALLY

Consider the two squares in figure 4-9, where the smaller one has a side of length 3 and the larger one has a side of length 4, and where the vertex (or corner) M of square $MNPQ$ is the center of square $ABCD$.

---

2. A triangle with two equal sides is called *isosceles*.
3. For a complete discussion of this, see A. S. Posamentier, *Advanced Euclidean Geometry* (Hoboken, NJ: John Wiley, 2002), pp. 13–15.

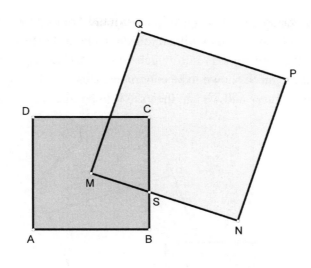

Figure 4-9

The question we raise is what is the area of the overlapping portion of the two squares? Does the exact relative position of the two squares matter? Since the exact placement of the two squares was not given—just the vertex *M*—suppose we rotate the larger square to the position shown in figure 4-10 or to that shown in figure 4-11. In either case, the "restrictions" mentioned in the original situation are still upheld, namely, that the vertex of one square is at the center of the other. From either of these new positions (figure 4-10 or 4-11), we can conclude that the overlapping portion is one-fourth of the area of the smaller square *ABCD*.

Yet this raises the question about whether this can assure us that this is not just an "extreme" case but also one that holds between the extreme or special positions.

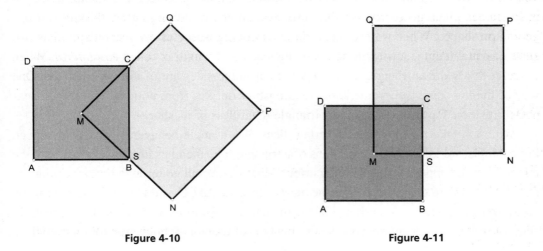

Figure 4-10   **Figure 4-11**

Suppose the larger square, *MNPQ*, is placed more "randomly," as shown in figure 4-12 and as it was positioned in our original question (figure 4-9). We can show (and, of course, prove) that the four regions into which square *ABCD* is partitioned all have the same area— since they can be shown to be congruent. This is a nice case of using extremes to guide us to a "conjecture" and we can then see it to be true for the random case shown in figure 4-12.

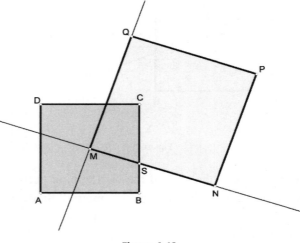

Figure 4-12

SOME TRIANGLE PROPERTIES—WITH MEDIANS

To impress you with the many wonderful, and often mind-boggling, relationships that exist in Euclidean plane geometry, we shall take you on an unfolding journey through various geometric shapes. When we refer to a triangle as being isosceles, we mean than it has two sides of equal length. A triangle with all sides of equal length is called *equilateral*. When we refer to a *scalene* triangle, it is one where all three sides are of different lengths. One way of finding some truly remarkable relationships derives from noticing that these unexpected relationships can occur in *any* triangle regardless of its shape!

We often take for granted that certain lines of a triangle behave as we would expect. For example, we know that the medians of a triangle (i.e., the lines joining a vertex with the midpoint of the opposite side) are concurrent—that is, they all go through the same point. In figure 4-13, *AF*, *BD*, and *CE* are the medians of triangle △*ABC*, and *G* is their point of intersection (or the point of concurrency of the medians). This, in itself, is a fascinating reationship, since it is completely independent of the type of triangle. It holds true for *all* triangles!

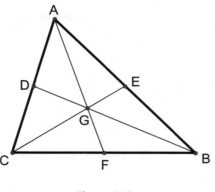

Figure 4-13

Furthermore, the medians partition the triangle into six smaller triangles, which happen to be equal in area. Again, this is true regardless of the shape of the triangle. That is, $Area\triangle ADG = Area\triangle AEG = Area\triangle BEG = Area\triangle BFG = Area\triangle CFG = Area\triangle CDG$.

We can also partition any triangle, say, $\triangle ABC$, into four congruent triangles ($\triangle ADE \cong \triangle CDF \cong \triangle FEB \cong \triangle EFD$) by using the midpoints of the three sides and joining them to form a medial triangle (see figure 4-14). Here, $\triangle EFD$ is called the *medial triangle* of $\triangle ABC$. Remember, these partitions hold true for *any* shape triangle.

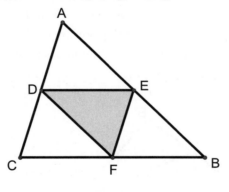

Figure 4-14

In other words, the area of triangle $\triangle ABC$ is four times the area of triangle $\triangle EFD$. That is, this relationship is a constant one that holds for all triangles.

In addition, point G (the point of intersection of the medians) is the center of gravity of the triangle, and is called the *centroid*. That is, if you want to balance a cardboard triangle on the tip of a pencil, you would need to find the centroid to place the point appropriately. We shall revisit the center of gravity a bit later. In the meantime, we can also appreciate another fascinating fact about the centroid of a triangle; namely, the centroid trisects each of the medians. In other words, in figure 4-15: $GD = \frac{1}{3}BD$, $GE = \frac{1}{3}CE$, and $GF = \frac{1}{3}AF$.

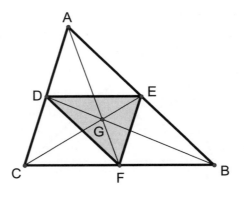

Figure 4-15

SOME QUADRILATERAL PROPERTIES

We shall now shift to inspect the quadrilateral and the midpoints of its sides. If we consecutively connect the midpoints of any quadrilateral with line segments, we always get a parallelogram (figure 4-16), which is called a *Varignon parallelogram*.[4] This is truly amazing, since it will hold true for any shape of a quadrilateral. Moreover, the Varignon parallelogram has half the area of the original quadrilateral, and its perimeter is equal to the sum of the lengths of the diagonals of the original quadrilateral. That is, in figure 4-16: the area of parallelogram *EFGH* is one-half of the area of quadrilateral *ABCD*, and the perimeter of parallelogram *EFGH* equals *AC + BD*.

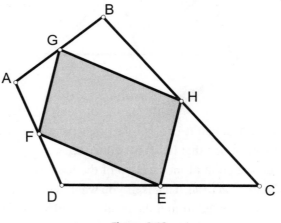

Figure 4-16

---

4. Named for Pierre Varignon (1654–1722), a French mathematician. Although discovered in 1713, it was first published (posthumously) in 1731.

Just for curiosity's sake, let's see what happens when we reduce the length of one side of the quadrilateral, say, *AB*, to length zero. We get a triangle with one vertex having points *A*, *B*, and *G* coincide as in figure 4-17. Here, too, the resulting Varignon parallelogram still exists and is one-half the area of the original triangle (or quadrilateral with a zero-length side). You might want to relate this to our result in figure 4-14, where we found that the triangle *ABC* was partitioned into four triangles of equal area, leaving us with a similar result when we consider the parallelogram *ADFE* in figure 4-14, which is analogous to parallelogram *AFEH* in figure 4-17. The perimeter relationship (perimeter of parallelogram *EFGH* equals *AC + BD*) also holds true for the situation shown in figure 4-17. It is not only nice to have this consistency, it is essential!

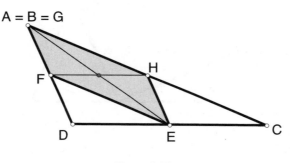

Figure 4-17

Of course, some quadrilaterals—when the midpoints of their sides are connected—will yield a special parallelogram, such as a rectangle, a square, or a rhombus (a parallelogram with all sides equal).

For example, suppose we consider a quadrilateral whose diagonals are perpendicular (see figure 4-18). When the side-midpoints of this quadrilateral are connected, a rectangle, *EFGH,* is produced.

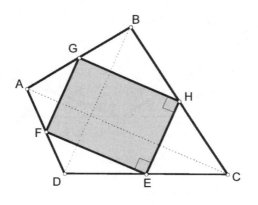

Figure 4-18

If the quadrilateral has perpendicular diagonals that are also the same length, then the quadrilateral produced by joining the side-midpoints is a special kind of rectangle that we know as a square! (See figure 4-19.)

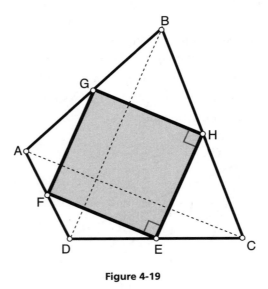

Figure 4-19

If the original quadrilateral has diagonals of equal length that are not perpendicular, then the quadrilateral formed by joining the side-midpoints will be a rhombus, as in figure 4-20.

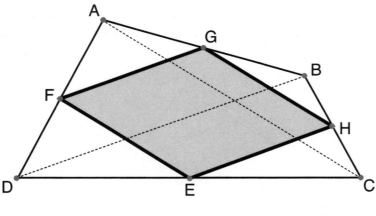

Figure 4-20

Naturally, we can continue this process of connecting consecutively the midpoints of the resulting parallelograms to get the series of parallelograms shown in figure 4-21.

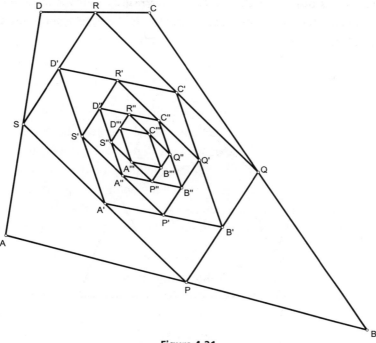

Figure 4-21

One of the prettier results from this investigation occurs when we can generate many special quadrilaterals from a randomly drawn quadrilateral. In the randomly drawn quadrilateral *ABCD* shown in figure 4-22 we connected the midpoints to form a parallelogram
(*MNOP*). We then found the intersections of the bisectors of the angles of parallelogram
MNOP to create rectangle *EFGH*. By bisecting each of the angles of this rectangle, we created the square *JKLR*. Therefore, from the randomly drawn quadrilateral *ABCD* we have
created three special quadrilaterals: a parallelogram, a rectangle, and a square.

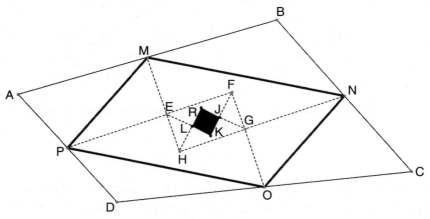

Figure 4-22

This is another example of the beauty of the geometric relationships that too often remain hidden. Their exposure is truly enlightening! You might want to try other ways of creating a square from a randomly drawn quadrilateral.

CENTER OF GRAVITY

Earlier (page 153), we located the center of gravity of a triangle (called the centroid) by finding the point of intersection of its medians. Some countries have run contests to find the "center" of their country. As you can see from figure 4-23, the contest submissions in Austria took on many creative methods. Yet the one that won the contest was for the town of Bad Aussee, where residents balanced a wooden map of the country on a point that was their town's location. Figure 4-23 is a translation of a plaque that designates the center of Austria in the town of Bad Aussee—as shown in the adjoining map (figure 4-24).

BAD AUSSEE
The geographic center in Austria
Latitude 47° 36' – Longitude 13° 47' east of border
(Dipl.-Ing. Otto Kloiber, Vienna)

Why can Bad Aussee name itself the "Center of Austria?" Here is a brief review: In 1949 the magazine *Die Große Österreich Illustrierte* had a competition with the challenge of finding the center of Austria. Each Austrian had been invited to take part. The "winner" received a free vacation, which was a luxury then. The participants tried with different methods to find the navel of Austria. For example, the following:

THE DIAGONAL-SOLUTION:
Find the four points on the border, draw a rectangular of latitude and longitude along those points. The intersection of the diagonals would be the center.

THE CORNER-METHOD:
Find the point which is, in total length, the furthest away from the borders of Austria. The geometric solution would be the center of a circle whose circumference is tangent to the edges of other countries bordering Austria, and which reaches furthest into Austrian territory.

THE FOCAL POINT-THEORY:
Glue the map of Austria onto cardboard and cut out the map exactly along the borders.

Put this Austria on the tip of a needle and move it around till the map balances.

The people of Bad Aussee started measuring and balancing and wrote to the magazine: "We are the ones!" The magazine confirmed (after a survey by the University of Vienna): "Bad Aussee is the center of Austria." The official ceremony with the actress Maria Andergast was on September 24, 1949.

In 1989 the local chamber of commerce of Bad Aussee used the 40th anniversary of the center-designation to celebrate with various festivities that the town is now the heart of Austria. As a permanent monument to remind everyone, that Bad Aussee is the center of Austria, a "center-stone" has been dedicated in the spa-gardens of the town.

The bronze plaque on the center-stone, a so-called "Omphalos" (Greek for "navel"), symbolizes the center. The inscription is in Gothic print and originated from the municipal seal of Aussee from 1505 and reads "Ause."

Furthermore, the local chamber of commerce strove to protect this title and applied to the Austrian patent office, which rendered this official designation in September 1991.

Figure 4-23

Figure 4-24

The exact center of the contiguous (lower) forty-eight states of the United States can be similarly determined by placing a cardboard map on a point to balance it. The point at which the map would be perfectly balanced is latitude 39 degrees 50' N, longitude 98 degrees 35' W, which is near the town of Lebanon in Smith County, Kansas (figure 4-25).

Figure 4-25

The balancing point is not always easy to locate without using "trial-and-error" methods. This is especially true for a quadrilateral. Finding the center of gravity of a quadrilateral is much more complicated and is by no means as neat as finding the centroid of a triangle. Yet, not wanting to create a very confusing diagram, it is merely a combination of finding the centroids of various triangles in the quadrilateral and then "combining" these points. Although it is not pretty to look at, we will show you how it is done so you can see the connection of the quadrilateral and the triangle.

This point may be found in the following way: Let L and N be the centroids of $\triangle ABC$ and $\triangle ADC$, respectively, in figure 4-26. Let K and M be the centroids of $\triangle ABD$ and $\triangle BCD$, respectively. The point of intersection, G, of LN and KM is the centroid of the quadrilateral $ABCD$.

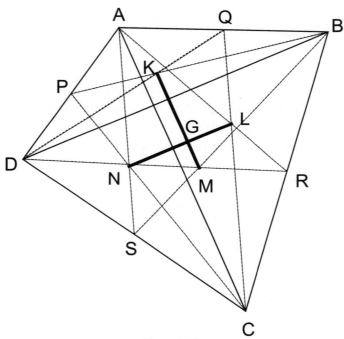

Figure 4-26

If you can get past the complexity of the diagram, you will see that we have simply located the centroids of the four triangles and then found the intersection of the two line segments joining them. We have then produced the analog of the triangle's centroid, or center of gravity—the point at which you could balance a cardboard quadrilateral. The centroid of a rectangle just happens to be much simpler to find. It is where you would expect it: at the point of intersection of the diagonals.

Let us return to the Varignon parallelogram *FJEH* (shown in figure 4-27). It may be more interesting to consider the *center point* (not the centroid) of a quadrilateral, that is, the point of intersection of the two segments joining the midpoints of the opposite sides of the quadrilateral. *G* is the center point of quadrilateral *ABCD*.

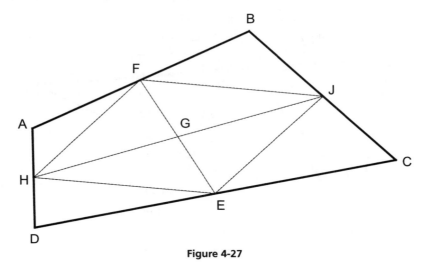

Figure 4-27

What is nice about these two line segments, *EF* and *HJ*, is that they bisect each other—that is, *G* is the midpoint of each of the segments *EF* and *HJ*. Furthermore, if we inspect the diagram carefully (but now look at figure 4-28), you will see that the segment joining the midpoints (*M* and *N*) of the diagonals (*BD* and *AC*) of a quadrilateral is bisected by the center point. That is, *G* bisects *MN*, or *MG* = *NG*. This is truly remarkable, especially if you bear in mind that we began with just *any* randomly drawn quadrilateral.

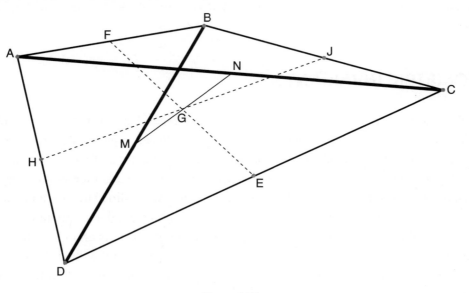

Figure 4-28

This suggests still another parallelogram in this figure. In figure 4-29, since *MN* and *FE* bisect each other, we can conclude that quadrilateral *FNEM* is also a parallelogram.

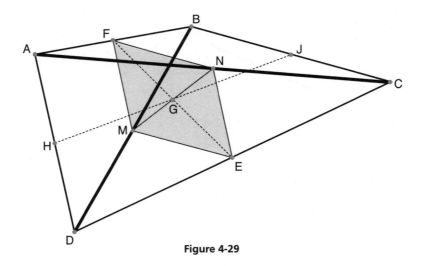

Figure 4-29

With all these remarkable relationships, many of which we would never have expected to realize from just a randomly drawn quadrilateral, we could tie these phenomena together further, if we could relate these two pairs of interior line segments: *AC* and *BD* with *EF* and *HJ*. Yes, they, too, can be connected with the following relationship: The sum of the squares of the lengths of the diagonals of *any* quadrilateral equals twice the sum of the squares of the lengths of the two segments joining the midpoints of the opposite sides of the quadrilateral. Here, for the quadrilateral *ABCD* (figure 4-29), we have, without proof:

$$(BD)^2 + (AC)^2 = 2\left[(EF)^2 + (HJ)^2 \right].$$

GEOMETRIC INVARIANTS

It is fascinating to notice how there are geometric figures that will remain constant even when other parts of a figure change. These are often referred to as *invariants* in geometry. We noticed one such example when we connected the midpoints of any quadrilateral and found that the resulting figure was always a parallelogram. The parallelogram—regardless of size or shape—remained a parallelogram, even when we modified the original quadrilateral. Another very nice example of this can be seen with an equilateral triangle. Consider the equilateral triangle $\triangle ABC$ in figure 4-30, where any point, *P*, is chosen in the triangle. Then, from the point *P*, perpendicular lines are drawn to each of the three sides of the triangle. We then measure the lengths of these three line segments, *DP*, *EP*, and *FP*, and find that their sum is 5.89 cm.

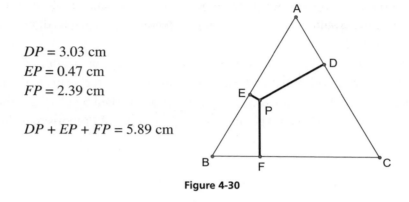

DP = 3.03 cm

EP = 0.47 cm

FP = 2.39 cm

DP + EP + FP = 5.89 cm

Figure 4-30

We now randomly move point *P* to another location in the same (equilateral) triangle *ABC* and find that the sum of the lengths of the three line segments *DP*, *EP*, and *FP* is once again the same as before (figure 4-31).

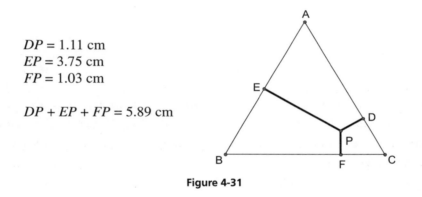

DP = 1.11 cm

EP = 3.75 cm

FP = 1.03 cm

DP + EP + FP = 5.89 cm

Figure 4-31

Repeating this process for yet another randomly selected location for point *P* in the same triangle *ABC*, in figure 4-32 we find that the sum of the lengths of the three line segments *DP*, *EP*, and *FP* is again the same.

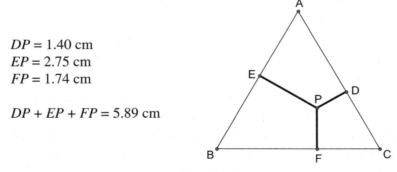

DP = 1.40 cm

EP = 2.75 cm

FP = 1.74 cm

DP + EP + FP = 5.89 cm

Figure 4-32

It appears that we may safely conclude that no matter where we place point P in the equilateral triangle, the sum of the lengths of the perpendicular segments drawn to the three sides will always be the same.

If we revisit figure 4-32, but this time show the altitude, or height, of (equilateral) triangle $\triangle ABC$, we will notice that its length, 5.89 cm, is the same as the sum of the three perpendicular segments (see figure 4-33).[5] This could be justified by placing point P at the vertex, A, where EP and DP would be of length 0, and FP would become the altitude AH.

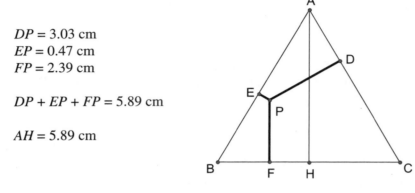

$DP = 3.03$ cm
$EP = 0.47$ cm
$FP = 2.39$ cm

$DP + EP + FP = 5.89$ cm

$AH = 5.89$ cm

Figure 4-33

In the previous examples we always chose point P to be located *in* $\triangle ABC$. Suppose we place point P *on* one of the sides of the triangle (we could say that points F and P coincide, as in figure 4-34). We notice that, once again, the sum of the perpendicular segments to the remaining two sides is still the same, 5.89 cm.

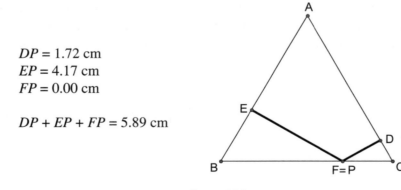

$DP = 1.72$ cm
$EP = 4.17$ cm
$FP = 0.00$ cm

$DP + EP + FP = 5.89$ cm

Figure 4-34

---

5. This is known as Viviani's Theorem, named after Vincenzo Viviani (1622–1703), an Italian mathematician, who was a student of Galileo Galilei (1564–1642). He published it in 1659 as a part of his work "De maximis et minimis divinatio in quintum Conicorum Appollonii Pergaei." Simply stated, the constant $DP + EP + FP = AH$ in figure 4-33.

Now, suppose we place point *P* on two sides of triangle Δ*ABC*. That is, we would then place this point at one of the vertices, which we did earlier—making the points *P*, *F*, *D*, and *C* coincide (figure 4-35).

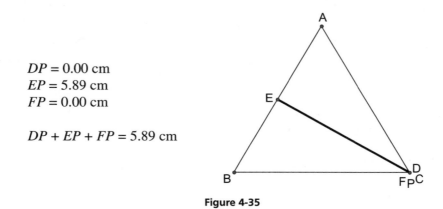

$DP = 0.00$ cm
$EP = 5.89$ cm
$FP = 0.00$ cm

$DP + EP + FP = 5.89$ cm

Figure 4-35

Once again, we notice (in figure 4-35) that the sum of the distances of the three perpendiculars—two of which are now of length zero—is still 5.89 cm. This would be a constant, namely, the altitude of the equilateral triangle Δ*ABC*.

From our observations of the sum of the distances of point *P* from each of the three sides of triangle Δ*ABC*, we find that whenever *P* is in or on the triangle, the sum is always the same, namely, the length of the altitude of the triangle. (Remember that the three altitudes of an equilateral triangle are equal.)

This idea can be extended to the isosceles triangle by taking a random point on the base and inspecting the sum of the distances (i.e., always the perpendicular distances) to the two equal sides of the triangle. We will find that no matter where the point is selected along the base (including at the endpoints, or vertices), the sum of the distances to the sides is the same (see figures 4-36 and 4-37).

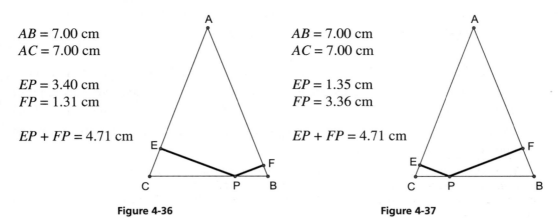

$AB = 7.00$ cm
$AC = 7.00$ cm

$EP = 3.40$ cm
$FP = 1.31$ cm

$EP + FP = 4.71$ cm

Figure 4-36

$AB = 7.00$ cm
$AC = 7.00$ cm

$EP = 1.35$ cm
$FP = 3.36$ cm

$EP + FP = 4.71$ cm

Figure 4-37

This sum can be found to be the same for all such points on base BC. We can even select the position of point P to be at one of the vertices of the triangle, say, point B. In figure 4-38, we notice that the sum of the perpendiculars is now the same as the altitude from vertex B to side AC, namely, 4.71 cm.

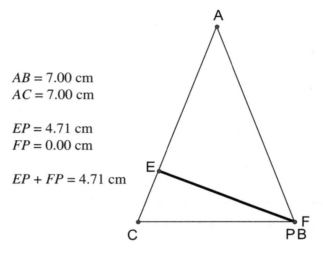

$AB = 7.00$ cm
$AC = 7.00$ cm

$EP = 4.71$ cm
$FP = 0.00$ cm

$EP + FP = 4.71$ cm

Figure 4-38

These cases of invariants pop up frequently in geometry and each time leave us with a feeling of awe. Just to review:

1. The sum of the perpendicular segments to each of the sides of an equilateral triangle from a given point in—or on—the triangle is a constant, namely, the length of the altitude of the triangle.
2. The sum of the perpendicular segments from any point on the base of an isosceles triangle to each of the two equal sides is a constant, namely, the altitude from a base angle vertex to the opposite side.

We have demonstrated the intrigue of invariants with the hope that you will be motivated to search for other such phenomena. Yet, to convince you of their veracity, a proof would be needed. Such proofs are relatively simple. To not detract from the "flow" of the discussion here, we will simply refer you to some books where several proofs of this can be found (see the bibliography).

NAPOLEON'S TRIANGLE

Although the amazing geometric phenomenon we are about to present is attributed to Napoleon Bonaparte (1769–1821), some critics assert that the theorem was actually discovered by one of the many mathematicians[6] with whom Napoleon liked to interact.

Simply stated, we begin our exploration of this geometric novelty with a randomly selected triangle—that is, here we choose one that has all sides of different lengths. We draw an equilateral triangle on each of the sides of this triangle (figure 4-39).

Next, we will draw line segments joining the remote vertex of each equilateral triangle with the opposite vertex of the original triangle (figure 4-40).

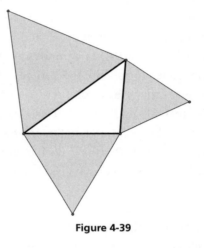

Figure 4-39

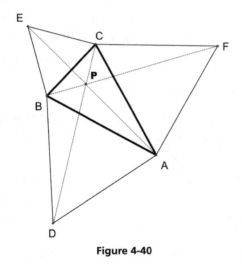

Figure 4-40

There are two important properties that evolve here—and are not to be taken for granted—namely, the three line segments we just drew are concurrent (that is, they contain a common point) and they are of equal length. Remember, this is true for a randomly selected triangle, which implies it is true for all triangles. That's the amazing part of this relationship.

---

6. Any of the following might have "helped" Napoleon arrive at this relationship:
Jean-Victor Poncelet (1788–1867), who served as one of Napoleon's military engineers and who later became one of the founders of projective geometry;
Gaspard Monge (1746–1818), a technical advisor to Napoleon who participated in the Egyptian campaign;
Joseph-Louis Lagrange (1736–1813), a French mathematician;
Lorenzo Mascheroni (1750–1800), who participated in the Italian campaign;
Jean Baptiste Joseph de Fourier (1768–1830), who participated in the Egyptian campaign; or
Pierre-Simon Marquis de Laplace (1749–1827), who was Napoleon's teacher (1784–85) and later served for six weeks as Napoleon's minister of the interior.

Furthermore, of all the infinitely many points in the original triangle, the point of concurrency is the point from which the sum of the distances to the three vertices of the original triangle is the shortest.[7] That is, in figure 4-41, from the point P, the sum of the distances to the vertices, A, B, and C (that is, $PA + PB + PC$) is a minimum. Also, the angles formed by the vertices of the original triangle at point P are equal. In figure 4-41, $\angle APB = \angle APC = \angle BPC$ ($=$ 120°). This point, P, is called the *Fermat point*, named after the French mathematician Pierre de Fermat (1607–1665).

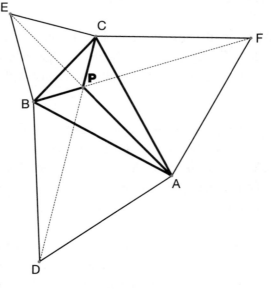

Figure 4-41

Now to Napoleon's triangle: When we join the center points of the three equilateral triangles (i.e., the point of intersection of the medians, angle bisectors, and altitudes), we obtain another equilateral triangle—called the *outer Napoleon triangle*. In figure 4-42 it is $\triangle KMN$.

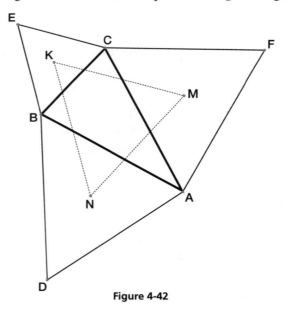

Figure 4-42

---

7. This is based on a triangle that has no angle greater than 120°. If the triangle has an angle greater than 120°, then the desired point is the vertex of the obtuse angle.

Had the three equilateral triangles—on each side of the randomly selected triangle—been drawn overlapping the original randomly selected triangle, then the center points of the three equilateral triangles would still produce an equilateral triangle. This is depicted in figure 4-43 as $\Delta K'M'N'$. It is called the *inner Napoleon triangle*.

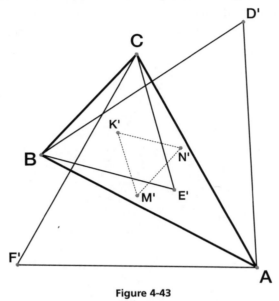

Figure 4-43

Now something comes up that may be a bit hard to picture. Let's consider the two Napoleon triangles in the same diagram (figure 4-44), which can be more easily seen if we omit the equilateral triangles we added to our original randomly selected triangle, as shown in figure 4-45.

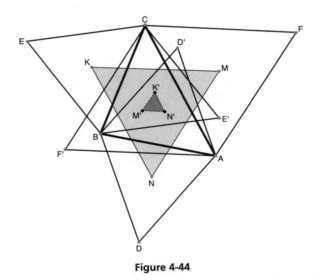

Figure 4-44

Consider the areas of the three triangles in the figure 4-45. The difference of the areas of the two Napoleon triangles (the inner and the outer) is equal to the area of the original randomly selected triangle. Thus, *Area △KMN – Area △K'M'N' = Area △ABC*. (See the bibliography.)

Again, we remind you that what makes this so spectacular is that it is true for an original triangle of *any* shape, which we tried to dramatize by using a scalene triangle that does not have any special properties.

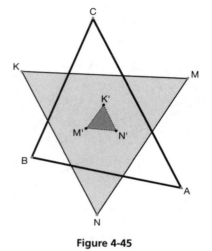

Figure 4-45

When we view the original triangle and the Napoleon triangles, they share the same centroid—that is, the point of intersection of the medians, which is also the center of gravity of the triangle. (See figure 4-46, where we show only the outer Napoleon triangle for clarity, but it is also true for the inner Napoleon triangle.)

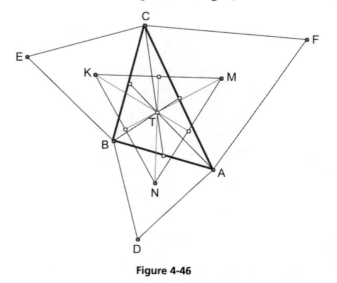

Figure 4-46

There are more surprising relationships that can be found on this Napoleon triangle. These were derived long after Napoleon (as he claimed) discovered the very basics of the equilateral triangle that evolved from the random triangle. Take, for example, the lines joining each vertex of the outer Napoleon triangle[8] to the vertex of the corresponding equilateral triangle drawn on each side of the original randomly selected triangle. First, we can show that these three lines are concurrent. That is, *DN*, *EK*, and *FM* meet at the common point *O*. Second, this point of concurrency, *O*, is the center of the circumscribed circle of the original triangle (figure 4-47).

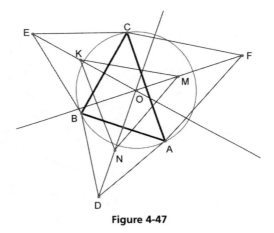

Figure 4-47

Believe it or not, there is even another concurrency in this figure. If we consider the lines joining a vertex of the Napoleon triangle with the remote vertex of the original triangle, we again get a concurrency of the three lines. In figure 4-48, *AK*, *BM*, and *CN* are just such lines and are concurrent.

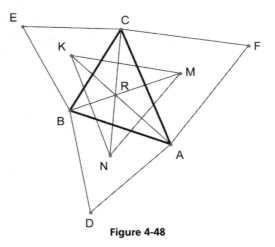

Figure 4-48

---

8. We shall use the outer one, just for clarity, but it would hold true for the inner one as well.

You may, by this point, get the feeling that just about anytime you have three "related" lines in a triangle, they must be concurrent. Well, to appreciate the concurrencies we have just presented, suffice it to say, concurrencies are not common. They may be considered exceptional when they occur. So that ought to make you really appreciate them!

As an extension of the equilateral triangles that we drew on the sides of the random triangle, we will now draw similar triangles (appropriately oriented) on each of the sides of a random triangle. In figure 4-49, we have the similarity[9] relationship:

$$\Delta AFB \sim \Delta BDC \sim \Delta CEA.$$

Both ΔDEF and ΔABC have the same centroid—the center of gravity, or the point of intersection of its medians. Notice that the three medians of ΔDEF, DR, EM, and FN, have G as their common point of intersection; while the medians of ΔABC, AY, BX, and CZ, also have G as their common point of intersection. The noteworthy aspect here is that we began with a random triangle and then constructed similar triangles of any shape on each of the three sides and, despite all that, the two triangles identified have a common centroid, G.

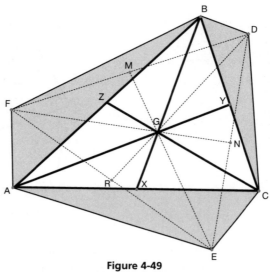

Figure 4-49

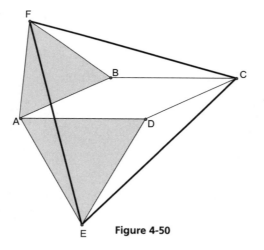

Figure 4-50

We can get an equilateral triangle by drawing equilateral triangles on two adjacent sides of any parallelogram. In figure 4-50, equilateral triangles ABF and ADE are drawn on the adjacent sides of parallelogram $ABCD$. We can show that triangle ECF is equilateral[10]—and this is true for all parallelograms, which is what makes this result so impressive.

9. Polygons are said to be *similar* if they have the same shape but not necessarily the same size.
10. Since triangles AFE, DCE, and BFC are congruent.

SQUARES ON THE SIDES OF POLYGONS

Now that we placed triangles on the sides of a random triangle and found many amazing relationships, let's see what we can find if we draw squares on the sides of a random triangle. In figure 4-51, we shall begin with a randomly drawn triangle, $\triangle ABC$, and then draw squares on sides AB, BC, and AC. This enables us to create triangles ADE, BKL, and CFH. We can show that each of these three triangles is equal in area to the original triangle ABC.

To do this we simply recall the trigonometric formula for the area of a triangle, namely, $Area = \frac{1}{2}\,ab\,\sin\angle C$. We notice that each of the angles of $\triangle ABC$ is supplementary to the angle sharing its vertex of the three triangles in question. That is, $\angle BAC + \angle DAE = 180°$, since the remaining angles around point A are two right angles whose sum is also $180°$. Because they are the sides of a square, the adjacent sides of the angles $\angle BAC$ and $\angle DAE$, $AB = AD$ and $AC = AE$. Therefore, $Area\triangle ADE = \frac{1}{2}\,(AD)(AE)\sin\angle DAE$ and $Area\triangle ABC = \frac{1}{2}\,(AB)(AC)\sin\angle BAC$. However, since $\sin\angle BAC = \sin(180° - \angle BAC) = \sin\angle DAE$, $Area\triangle ABC = Area\triangle ADE$. This same reasoning will hold for the other two triangles: $\triangle BLK$ and $\triangle CFH$.

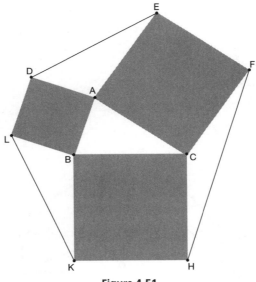

Figure 4-51

There is another hidden gem in this configuration. If, as in figure 4-52, we draw squares on the sides of the triangles DAE, LBK, and FCH, we can create quadrilaterals $DLTU$, $KHQS$, and $EFNR$. Amazingly, each of these three quadrilaterals has five times the area of the original $\triangle ABC$. The motivated reader may wish to prove this.[11]

11. A "proof without words" can be found in M. N. Deshpande, "Proof without Words beyond Extriangles," *Mathematics Magazine* 82, no. 3 (June 2009): 208.

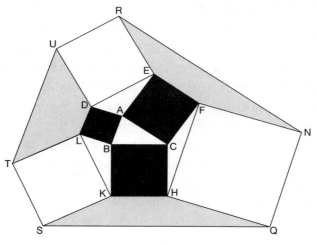

Figure 4-52

In figure 4-53 we once again construct squares on the sides of triangle *ABC*. Notice that the circumcircles[12] of the two squares *ABEF* and *ACJD* meet at point *P*. When we construct the circle *BC* as diameter, we find that it contains the point *P*. Needless to say, this is quite spectacular, and at the same time, not easy to discover. As an ongoing challenge, you might try to find other relationships in these figures.

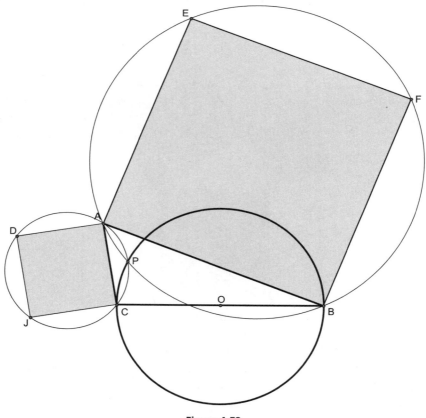

Figure 4-53

---

12. The *circumcircle* of a polygon is a circle where all vertices of the polygon lie on the circle.

We can also place squares on the sides of *any* parallelogram. Again, we stress that this is a *randomly* selected parallelogram. This is shown in figure 4-54, where squares were drawn on the sides of parallelogram *ABCD*. We can locate the centers of each of the squares by getting the point of intersection of each square's diagonals. Joining these four points, amazingly, yields another square, *RQTS*. This theorem is called the Yaglom-Barlotti Theorem: [13]

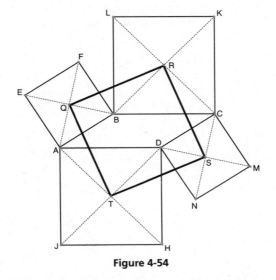

Figure 4-54

Moreover, the diagonals of this square (*RQTS*) meet at the same point (*P*) as the intersection of the diagonals of the original parallelogram (figure 4-55). To fully appreciate this unexpected relationship, you need to keep in mind that this will hold true for any parallelogram.

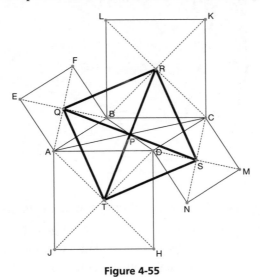

Figure 4-55

---

13. Named after the Russian mathematician Isaak Moisejewitsch Yaglom (1921–1955) and the Italian mathematician Adriano Barlotti (1923–).

We can also show that the sum of the areas of the four squares on the sides of the parallel-ogram is equal to the sum of the areas of the squares on the diagonals of the parallelogram. That is, in figure 4-56, *Area □ABFE + Area □AJHD + Area □MNDC + Area □BCKL = Area □BDVW + Area □ACYZ.*

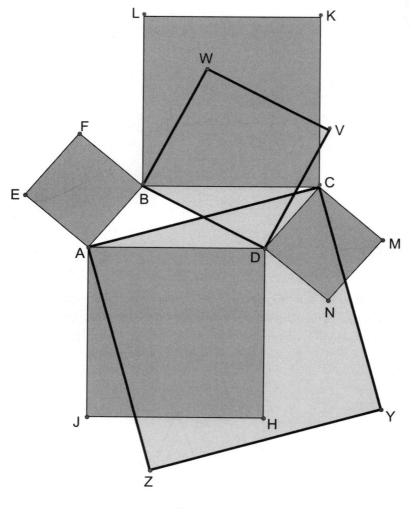

Figure 4-56

Suppose we now combine two of the relationships we established to this point: those related to the Napoleon triangle and those we just established by placing squares on the sides of a parallelogram. We will draw squares on sides of a randomly drawn quadrilateral as we show in figure 4-57. There you will notice that the line segments that join the centers of the four squares are equal and are perpendicular to each other. This was first published by the French engineer Edouard Collignon (1831–1913).

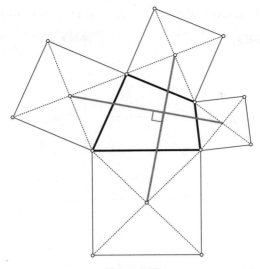

Figure 4-57

We already know what happens when we join the centers of the squares drawn (externally) on the sides of a parallelogram—we get a square (as in figure 4-55). It is interesting to see what kind of quadrilateral we would get if we joined the centers of the squares drawn (externally) on the sides of various other kinds of quadrilaterals, such as rectangles, squares, trapezoids, and kites.[14]

When the original quadrilateral *ABCD* is a rectangle, we get a square (*PQRS*); that is, when the centers of the side-squares are connected as in figure 4-58.

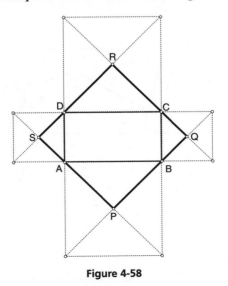

Figure 4-58

---

14. A *kite* is a quadrilateral with two disjoint pairs of congruent adjacent sides, in contrast to a parallelogram where the congruent sides are opposite.

When the original quadrilateral *ABCD* is an isosceles trapezoid, we get a kite (*PQRS*); that is, when the centers of the side-squares are connected as in figure 4-59.

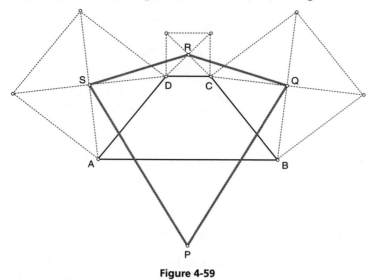

Figure 4-59

When the original quadrilateral *ABCD* is a kite, we get an isosceles trapezoid (*PQRS*); that is, when the centers of the side-squares are connected as in figure 4-60.

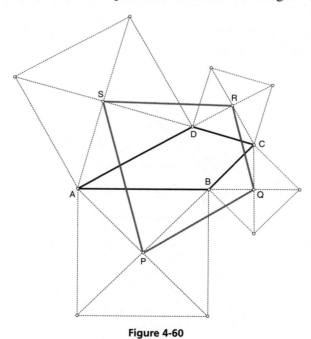

Figure 4-60

This mutual "relationship" between a kite and an isosceles trapezoid is one to be cherished.

Perhaps the most famous example of placing squares on the sides of a polygon, in this case a triangle, is when squares are drawn on the three sides of a right triangle.

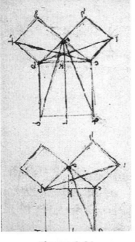

Figure 4-61

The picture shown in figure 4-61 is a sketch by Leonardo da Vinci (1452–1519) that demonstrates the Pythagorean Theorem and can be seen in many books dating back hundreds of years. It shows that the sum of the areas of the squares on the legs of a right triangle is equal to the area of the square on the hypotenuse of the right triangle. We can see this in figure 4-62; that is, $AreaADEC + AreaBCFH = AreaABJK$.

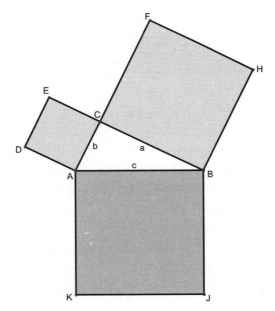

Figure 4-62

There are well over 520 different proofs[15] of this celebrated theorem, some of which were done by Pythagoras (ca. 570–510 BCE), Euclid (ca. 365–ca. 300 BCE), Leonardo da Vinci (1452–1519), Albert Einstein (1879–1955), and US president James A. Garfield (1831–1881)—when he was a member of the House of Representatives. Yet what most people remember about the theorem is that $a^2 + b^2 = c^2$, where the legs of the right triangle are of length a and b, while the length of the hypotenuse is c. This is analogous to the previous statement about the areas of the squares on the sides of the right triangle. Since the ratio of the areas of any similar polygons is the square of their ratio of similitude,[16] we shall now consider polygons other than squares on the sides of the right triangle. We could even use similar right triangles, as the similar polygons, to place on the sides of right triangles. By doing this, we can "visually" prove the Pythagorean Theorem in a manner that could be seen as intuitive.

Consider the right $\triangle ABC$ with right angle ACB and altitude CD. In figure 4-63, we cover each of the three similar right triangles, $\triangle ACD$, $\triangle CBD$, and $\triangle ABC$, with a congruent triangle. We can plainly see that $Area\triangle ACD + Area\triangle BCD = Area\triangle ABC$.

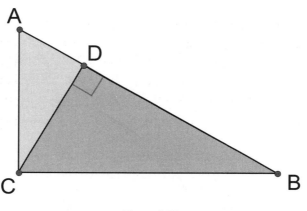

Figure 4-63

We then flip them over on their respective hypotenuses as is shown in figure 4-64, so that we can conclude that the sum of the areas of the similar right triangles ($\triangle ACD''$ and $\triangle BCD'$) on the legs of the right triangle $\triangle ABC$ is equal to the area of the similar right triangle ($\triangle ABC'$) on the hypotenuse. As we said earlier, since the ratio of the areas of any similar polygons is equal to the ratio of the squares of the corresponding sides, we get $(AC)^2 + (BC)^2 = (AB)^2$, which is the Pythagorean Theorem—this proves it!

---

15. The book *The Pythagorean Proposition*, by Elisha S. Loomis, contains 370 different proofs of the Pythagorean Theorem and was originally published in 1940 and republished by the National Council of Teacher of Mathematics in 1968.
16. The *ratio of similitude* of two similar polygons is the ratio of the lengths of any two corresponding sides.

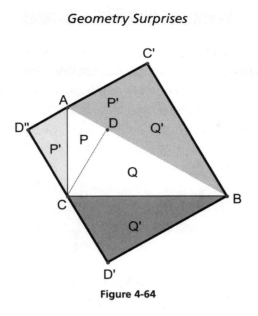

Figure 4-64

BISECTING AND TRISECTING ANGLES

A very astonishing result occurs when we bisect each angle of a *cyclic quadrilateral*, which is a quadrilateral that can be inscribed in a circle.[17] Figure 4-65 shows that if we bisect each of the angles of a cyclic quadrilateral (*ABCD*) and we mark the points where these bisectors intersect the circle, the four points will always determine a rectangle (*EFGH*). Imagine how amazing this is. Regardless of the shape of the cyclic quadrilateral, the four intersection points of the angle bisectors with the circumcircle will always determine a rectangle.

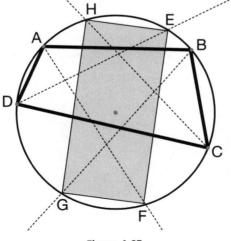

Figure 4-65

_____

17. A quadrilateral that can be inscribed in a circle is one where all four vertices lie on the same circle. Remember, all triangles can be inscribed in a circle, but not necessarily all quadrilaterals—only cyclic quadrilaterals.

It can be expected that one might ask: "Under what conditions would this rectangle (*EFGH*) be a square?" The answer is not exactly intuitive, but it is one that is quite satisfying. The rectangle would be a square when the diagonals of the cyclic quadrilateral are perpendicular. This is shown in figure 4-66.

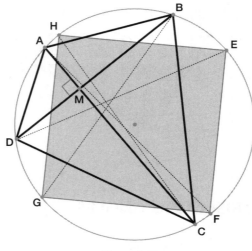

Figure 4-66

In geometry, investigations of relationships seem to never come to an end. We see that here, as there is another fascinating, and completely unexpected, relationship in this diagram; that is, if we focus on the four points of intersection of the angle bisectors (see figure 4-67). These bisectors meet in four points that lie on the same circle. We call these *concyclic points*, and, of course, we know that the four points determine a cyclic quadrilateral.

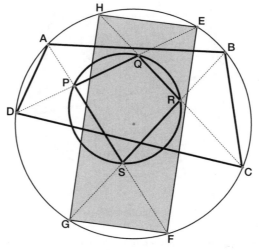

Figure 4-67

This unsurprising result should not to be taken for granted, since unlike three non-collinear points,[18] which always lie on a common circle, four points don't usually have this relationship. When they do lie on the same circle, it is something special to take note of. For example, if the four points are the vertices of a nonrectangular parallelogram, they will not be concyclic. Or put another way, the only parallelogram that can be inscribed in a circle is the rectangle (which, of course, includes the square as well).

We can look into this last relationship to see what might be generalized from it. Suppose we find the points of intersection of the angle bisectors of *any* quadrilateral (i.e., this time also including non-concyclic quadrilaterals). We will then conclude that they also determine a cyclic quadrilateral, as shown in figure 4-68. Here we notice that the original quadrilateral *ABCD* is *not* cyclic and yet the four intersection points of its angle bisectors determine a cyclic quadrilateral, *PQRS*.

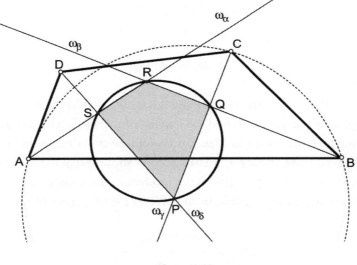

Figure 4-68

If the original quadrilateral is a parallelogram, then we notice (figure 4-69) that the angle bisectors meet to form a rectangle, which is always a cyclic quadrilateral.[19]

---

18. Three points that do not lie on the same line.
19. A cyclic quadrilateral always has supplementary opposite angles (i.e., their sum is 180°). Therefore, a rectangle is always a cyclic quadrilateral.

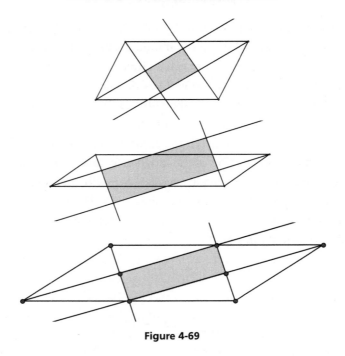

Figure 4-69

Perhaps one of the most famous facts about a cyclic quadrilateral is the relationship between the lengths of the diagonals and the lengths of the sides. The discovery of this remarkable relationship is attributed to Claudius Ptolemaeus of Alexandria (often referred to as Ptolemy, who lived from about 83 to 161 CE). He showed that the sum of the products of the pairs of opposite side lengths equals the product of the lengths of the diagonals. For figure 4-70, Ptolemy's relationship tells us that $AB \cdot CD + AD \cdot BC = AC \cdot BD$.

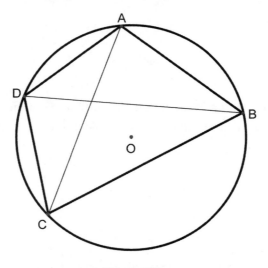

Figure 4-70

After this little sojourn through the realm of quadrilaterals, we shall revisit the triangle to admire even more astonishing facts about this omnipresent geometric shape.

MORLEY'S TRIANGLE

One of the most amazing theorems in Euclidean geometry was discovered by Frank Morley (1860–1937)[20] as recently as 1899. You might wonder why it took so many years of geometric exploration to eventually discover this heretofore-unknown relationship. This incredible geometric result involves trisecting an angle. For many years in antiquity, until almost modern times, mathematicians could not prove that a randomly selected angle could be trisected—that is, divided into three equal angles with the usual Euclidean tools: an unmarked straightedge and compasses. Of course, with other tools one could trisect an angle, but our study of Euclidean geometry restricts exact constructions to these two tools. In the nineteenth century, it was proved that one could not trisect an angle with these Euclidean tools.[21] This might have stifled any further investigations of geometric figures that involved trisecting an angle. However, of course we can consider the trisectors of an angle. When we do, we can discover an amazing relationship.

This relationship that today bears Morley's name states that the points of intersection of the adjacent angle trisectors of any triangle will always determine an equilateral triangle. This is a powerful statement, as it holds true for *any* shape of triangle. We show this in figures 4-71, 4-72, and 4-73 for a few commonly shaped triangles.

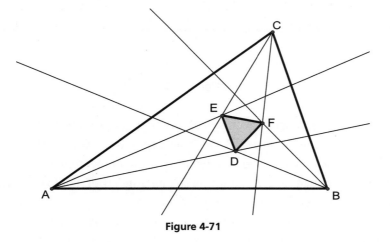

Figure 4-71

---

20. He was the father of the writer Christopher Morley (1890–1957).
21. First proved by the French mathematician Pierre Laurent Wantzel (1814–1884), "Recherches sur les moyens de reconnaître si un Problème de Géométrie peut se résoudre avec la règle et le compas," *Journal de Mathématiques Pures et Appliquées* 1, no. 2 (1837): 566–72.

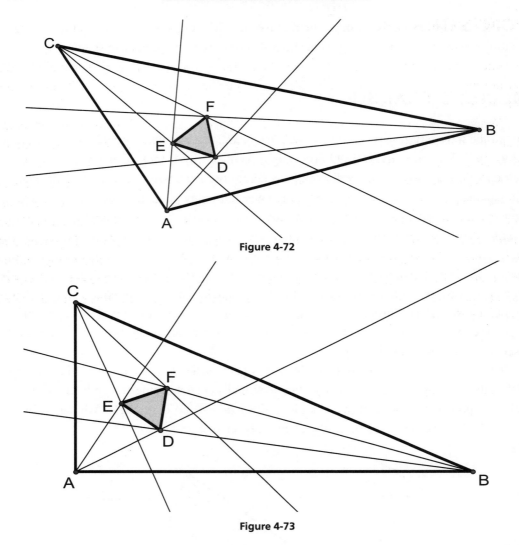

Figure 4-72

Figure 4-73

Remember that when we trisect the angles of *any* triangle, we always will create an equilateral triangle, as shown in figures 4-71, 4-72, and 4-73. Interestingly, this rather easy relationship to see could be considered one of the most difficult theorems to prove in Euclidean geometry. Yet we presented it here, not for its challenging proof, but to have you appreciate a truly phenomenal occurrence in plane Euclidean geometry.

At this point, you should have a fine idea about the many wonderful geometric relationships that are all too often not presented to high school students, thus depriving them of the opportunity to enhance and enrich their understanding and appreciation of geometry. But bear in mind, the beauty of the many geometric relationships we have seen thus far is that they involved randomly selected triangles or quadrilaterals.

POINTS ON A CIRCLE

We know that any three points that do not lie on the same line will lie on some circle. We could rephrase this to say the same thing in another way: any three non-collinear points determine a unique circle.

When we have four points on a circle—as we have done earlier for some special quadrilaterals, such as a rectangle—we can connect them consecutively and form a *cyclic* quadrilateral, one that we also can say is inscribed in a circle.

However, to find other predetermined points that lie on the same circle is not trivial —moreover, it is quite amazing when that does happen. Let's investigate just such a situation. One unexpected relationship was first published in 1765 by the Swiss mathematician Leonhard Euler (1707–1783), the most prolific mathematician of all time. He proved that the midpoints of the sides of any triangle and the feet of the altitudes[22] of that triangle all lie on the same circle. In figure 4-74, we can see that the three midpoints of the sides of $\triangle ABC$, namely, points D, E, and F, lie on the same circle as do the feet of the three altitudes (AK, BT, and CL), namely, the points K, T, and L.

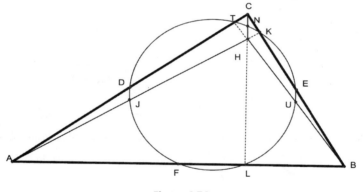

Figure 4-74

As if this were not spectacular enough, in 1820 another three points were placed on this very same circle when it was revealed in a journal article by the French mathematicians Charles-Julian Brianchon (1783–1864) and Jean-Victor Poncelet (1788–1867) that the midpoints of the segments connecting the vertices and the orthocenter,[23] which, in figures 4-74 and 4-75, are the points J, U, and N, lie on the circle. This is truly remarkable: nine points related to the same triangle all lying on the same circle! Hence, this circle is called the *nine-point circle* of a triangle. The German mathematician and teacher Karl Wilhelm Feuerbach (1800–1834) found four more points on this circle. Therefore, we also refer to

---

22. The foot of an altitude is the point at which the altitude of a triangle intersects the side of the triangle, which is then referred to as the base of the triangle. Two of these feet can lie outside the triangle.
23. The *orthocenter* is the point of intersection of the three altitudes of a triangle.

this circle as the *Feuerbach circle*, which is also tangent to the inscribed circle and the escribed circles (each of which is tangent to the three sides of the original triangle).

But we are not finished with this circle, since there are a number of other amazing facts to be mined from it.

Suppose we consider the circle circumscribed about the triangle—sometimes called the *circumcircle* of a triangle. We know that every triangle has a unique circumcircle. Focus on the line joining the center of the circumcircle with the orthocenter, which we call *HO* in figure 4-75. This is often referred to as the *Euler line*. By finding the midpoint, *M*, of the segment *HO* you have now also found the center of the nine-point circle (shown in figure 4-76).

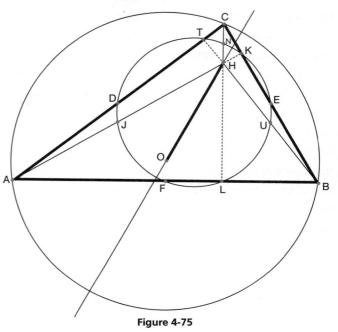

Figure 4-75

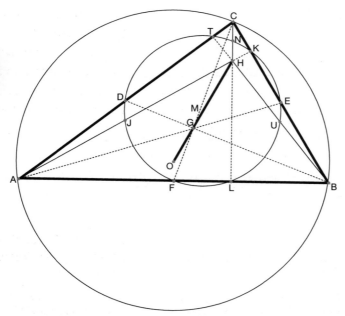

Figure 4-76

Recall that we found the centroid (or center of gravity) of a triangle by getting the point of intersection of the three medians of a triangle. Strangely enough, this point, the centroid, just happens also to be on the Euler line—situated exactly one-third from one end of the Euler line segment *HO*. In figure 4-76, the centroid is labeled as point *G*.

Incidentally, the radius of the nine-point circle is half the length of the radius of the circumcircle. That is, in figure 4-77, the length of *JM* (the radius of the nine-point circle) is one-half the length of *AO* (the radius of the circumcircle).

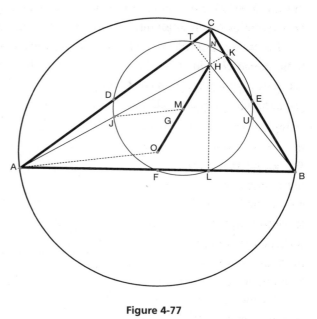

Figure 4-77

There are many more relationships that involve the nine-point circle, but they become a bit more cumbersome. We have extracted here the more spectacular relationships of this amazing circle for you to enjoy. For those eager to explore more such relationships related to the nine-point circle, we refer you to *Advanced Euclidean Geometry*.[24]

DUALITY

Many statements in geometry involve relationships between points and lines. Some new geometric relationships can be found by interchanging these key words. For example, when the word *point* is replaced by the word *line* and the word *line* is replaced by the word *point* each time they are used in the statement, we can get the *dual* of the original statement. (Occasionally, some other modifications may need to be made in order to make sense.) For example, we know that "any two points determine a line." The dual of this statement is that "any two lines (nonparallel and in the same plane) determine a point."

Other examples of duality are:

- Any *point* can contain an infinite number of *lines* through it.
- Any *line* contains an infinite number of *points* on it.

This principle of duality was discovered by Charles-Julien Brianchon (1783–1864) as he used this relationship on a theorem by the French mathematician Blaise Pascal (1623–

24. A. S. Posamentier, *Advanced Euclidean Geometry* (Hoboken, NJ: John Wiley and Sons, 2002). See also H. S. M. Coxeter and S. L. Greitzer, *Geometry Revisited*, New Mathematical Library 19 (Washington, DC: Mathematical Association of America, 1967).

1662). We shall look at each of these two spectacular relationships—both for their own beauty and for their application of the fascinating concept of duality.

Pascal discovered the following relationship: **If a hexagon with no pair of opposite sides parallel is inscribed in a circle, then the intersections of the opposite sides (extended, of course) lie on the same line.** In figure 4-78, the pairs of opposite sides of hexagon *ABCDEF* are (*AB, DE*), (*AF, CD*), and (*BC, EF*) and they meet at points *L*, *N*, and *M*, which you can see lie on the same line and therefore are said to be *collinear*.

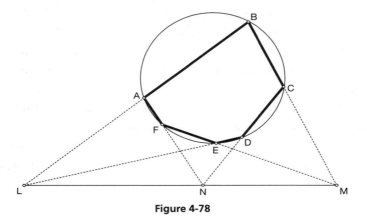

Figure 4-78

As a rather "wild" extension of Pascal's relationship, we can take the points *A*, *B*, *C*, *D*, *E*, and *F* as they are given to be the vertices of the hexagon and rearrange the order in which they are placed on a circle. As in figure 4-79, we take the previously defined "opposite sides," (*AB, DE*), (*AF, CD*), and (*BC, EF*), and determine their points of intersection: *L*, *N*, and *M*, and, amazingly, we find these points also to be collinear! In other words, even after rearranging the vertices on a circle, our previously defined "opposite sides" yielded collinear points of intersection. This is one of the beauties of mathematics!

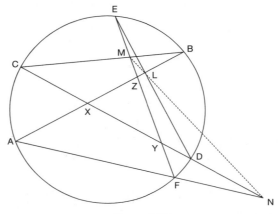

Figure 4-79

Let us now construct the dual of Pascal's relationship.

Pascal's Relationship	**The Dual of Pascal's Relationship**
The <u>points of intersection</u> of the opposite <u>sides</u> of a hexagon <u>inscribed in</u> a conic section[25] *are <u>collinear</u>.*	*The <u>lines joining</u> the opposite <u>vertices</u> of a hexagon <u>circumscribed about</u> a conic section are <u>concurrent</u>.*

In 1806, at the age of twenty-one, a student at the École Polytechnique, Charles-Julien Brianchon (1783–1864), published an article in the *Journal de L'École Polytechnique* that was to become one of the fundamental contributions to the study of conic sections in projective geometry. His development led to a restatement of the then somewhat-forgotten theorem of Pascal.

Brianchon's Theorem states: **In any hexagon circumscribed about a circle,**[26] **the three diagonals cross each other at a common point**.

In figure 4-80, (A, D), (C, F), and (B, E) are the pairs of opposite vertices. And we notice that the three diagonals (the segments joining the opposite vertices) meet at point P.

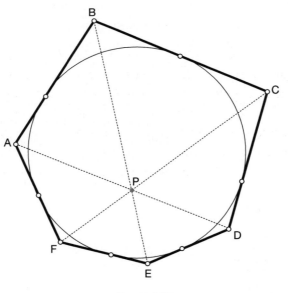

Figure 4-80

The concept of duality brings together many notions in mathematics and allows us to see connections that otherwise would not be so apparent.

---

25. A *conic section* is a plane figure formed by cutting a cone with a plane. Conic sections include circles, ellipses, parabolas, and hyperbolas.
26. Or actually any other conic section as well.

DESARGUES' RELATIONSHIP

We are about to present an unusual relationship that is its own dual. This time the relationship deals with the actual placement of geometric figures, which differs from previous relationships that depended only on shape and size, and where placement (or relative placement) was largely of no real concern to us.

During his lifetime, the French mathematician Gérard Desargues (1591–1661) did not enjoy the stature as an important mathematician that he attained in later years. This lack of recognition was in part due to the then-recent development of analytic geometry by René Descartes (1596–1650) and to his introduction of many new and largely unfamiliar terms. (Incidentally, we make every effort in this book not to introduce any unnecessary terms to keep it more reader-friendly. We want to learn from the Desargues experience.)

In 1648, Desargues' pupil, Abraham Bosse (ca. 1604–1676), a master engraver, published a book entitled *Manière universelle de M. des Argues, pour pratiquer la perspective par petit-pied comme le géométral* (we write his name today is Desargues), which was not popularized until about two centuries later. This book contained a theorem that, in the nineteenth century, became one of the fundamental propositions of the field of projective geometry. It is this theorem that is of interest to us here. It involves placing any two triangles in a position that will enable the three lines joining corresponding vertices to be concurrent. Remarkably enough, when this is achieved, the pairs of corresponding sides (extended) meet in three collinear points.

This relationship is shown in figure 4-81 and may be explained as follows:
If $\Delta A_1B_1C_1$ and $\Delta A_2B_2C_2$ are situated so that the lines joining the corresponding vertices, A_1A_2, B_1B_2, and C_1C_2, are concurrent, then the pairs of corresponding sides (extended) intersect in three collinear points (Desargues' Theorem).

In figure 4-81, lines A_1A_2, B_1B_2, and C_1C_2 all meet at P. Therefore, according to Desargues' Theorem, we can conclude that if lines B_2C_2 and B_1C_1 meet at A'; lines A_2C_2 and A_1C_1 meet at B'; and lines A_2B_2 and A_1B_1 meet at C', then A', B', and C' are collinear.

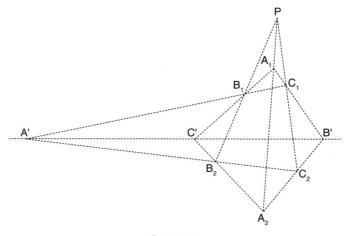

Figure 4-81

As we said earlier, this theorem is a self dual. Thus, the dual of this relationship is also true; namely, that if $\Delta A_1B_1C_1$ and $\Delta A_2B_2C_2$ are situated so that the pairs of corresponding sides (extended) intersect in three collinear points A', B', and C' (lines B_2C_2 and B_1C_1 meet at A'; lines A_2C_2 and A_1C_1 meet at B'; and lines A_2B_2 and A_1B_1 meet at C'), then the lines joining the corresponding vertices, A_1A_2, B_1B_2, and C_1C_2, are concurrent: again, at point P.

What makes this relationship remarkable is that it holds true for any shape triangles just as long as they can be placed in a position that will allow the concurrency or the collinearity.

SIMSON'S (WALLACE'S) RELATIONSHIP

One of the great injustices in the history of mathematics involves a concept originally published by William Wallace (1768–1843) in Thomas Leybourn's (1770–1840) *Mathematical Repository* (1799–1800), which, through careless misquotes, has been attributed to Robert Simson (1687–1768), a famous English interpreter of Euclid's *Elements*.

Robert Simson's geometry book, which was in print in Great Britain for well over a century, was very popular in the eighteenth and nineteenth centuries and is one of the bases for today's high school geometry courses. When Wallace's concept became known, it was soon being attributed to Robert Simson. This occurred since it was felt that a nonanalytic geometry proposition would not come from Descartes, so who else but Simson could have developed it? Well, this false attribution is still in vogue today, and the relationship is popularly called Simson's Theorem. It states that the feet of the perpendiculars drawn from any point on the circumcircle of a triangle to the sides of the triangle are collinear.

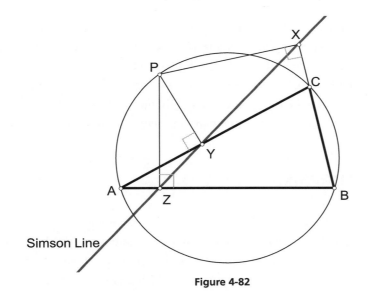

Figure 4-82

In figure 4-82, we select *any point* on the circumcircle of $\triangle ABC$. From that point, P, we draw the perpendicular lines to each of the three sides of the triangle, namely, PX, PY, and PZ to meet the sides BC, AC, and AB, at points X, Y, and Z, respectively. Regardless of the shape of the triangle or the placement of point P on the circumcircle, the three points X, Y, and Z will always lie on the same line (collinear). This line is often called the *Simson line*. This is a truly remarkable result, since it works for any shape triangle and any point on its circumscribed circle.

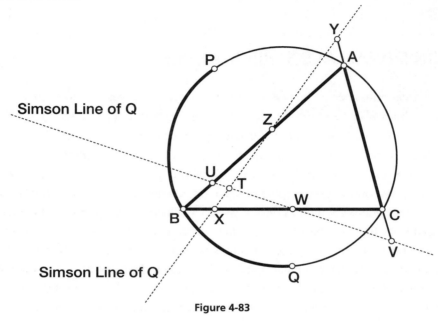

Figure 4-83

In figure 4-83 we have two Simson lines for triangle ABC determined by points P and Q. To keep the diagram from being too cluttered, we omitted the perpendicular line segments and only indicated the feet of these perpendiculars, which determine the two Simson lines. Amazingly, the angle ($\angle XTW$) formed by the two Simson lines is one-half the measure of the arc (QBP) between the two points P and Q. That is, $\angle XTW = \frac{1}{2}\, QBP$. There are more relationships involving the Simson line that you may want to discover for yourself.

A GEOMETRIC SURPRISE

There are times when geometric relationships are truly unanticipated—or even mind-boggling. Take, for example, the hypothetical situation where a 40,000 km rope is tied tightly along the equator of the earth, circumscribing the entire earth sphere (figure 4-84). Now suppose we lengthen this enormously long rope by 1 meter. It is then no longer tightly tied around the earth. If we lift this loose rope uniformly around the equator so that it is equally spaced above the equator, will a mouse fit beneath the rope?

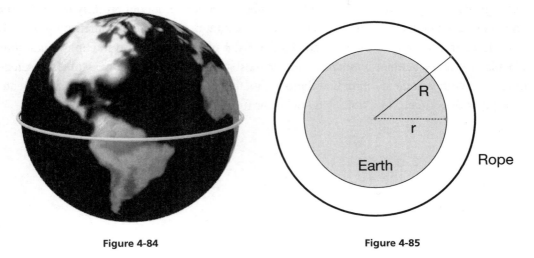

Figure 4-84   Figure 4-85

To analyze this situation, let's focus on figure 4-85, where we picture the two concentric circles: the rope and the circle of the earth. Our question requires us to determine the distance between the two circles. Suppose the sizes of the circles (figure 4-85) were not given. Let's see if we can make some generalization about the distance between the circles. Let us then assume (without loss of generality) that the small (inner) circle is extremely small, so small that it has a radius (r) and a circumference (C) of length 0, thus reducing the inner circle to a point. Then the distance between the circles is merely the radius (R) of the larger circle.

We can easily find the circumference of this larger circle using the well-known formula, and adding one meter: $2\pi R = C + 1$. When we shrink (theoretically, of course) the smaller circle—in this case, the earth—to a zero size (i.e., $C = 0$), then the circumference of the larger circle is $2\pi R = 0 + 1 = 1$. The distance between the circles—which is now just the radius of the larger circle—is $R = \frac{1}{2\pi} = 0.159$ meters. The same result would be found for any size inner circle.[27] Therefore, we can answer the question that 0.159 meters (or about 0.52 feet) would allow a mouse to comfortably fit beneath the rope. Imagine, by lengthening the rope around the earth by 1 meter, we have enough space beneath it to fit a cat! This should show you that even in geometry not everything is "intuitively obvious" (as we saw visually at the beginning of the chapter) and that there are geometric "facts" that can trick you intuitively as well as optically.

---

27. For the proof and more to this paradox see "A Rope around the Equator" and "A Rope around the Regular Polygons," in A. S. Posamentier and I. Lehmann, π: *A Biography of the World's Most Mysterious Number* (Amherst, NY: Prometheus Books, 2004).

At this point, you should have a fine idea about the many geometric relationships that all too often have not been presented in high school, when the study of geometry took center stage. This might have deprived a fair number of people of an opportunity to genuinely appreciate the beauty of geometry. The many amazing properties of geometric figures we considered in this chapter take on the characteristic of beauty when we keep in mind that we used ordinary geometric shapes—oftentimes stressing the randomness of their selection. Take with you the thought that there are many fascinating relationships that permeate all aspects of geometry—a field wide open for further explorations!

Chapter 5

MATHEMATICAL NUGGETS: AMAZING, BUT TRUE!

Calculating the chance of an event happening is done on a regular basis throughout our everyday lives. Most of the time, we find the results predictable. For example, if we flip a coin one hundred times, we would expect to get fifty heads and fifty tails—or at least something close to that. If we select at random twelve cards from a deck of fifty-two cards, we would expect one-quarter of them, or three, to be spades—or at least close to that. Yet there are some probabilities that defy our intuition. We shall examine some of these surprising and quite unexpected results here. Many of these seemingly common situations have intuition-defying results. We hope your initial frustration caused by calling into question your intuition will eventually be replaced by the joy found in enlightenment.

BIRTHDAY MATCHES

Here we present one of the most surprising results in mathematics. It is one of the best ways to convince the uninitiated of the "power" of probability. Aside from being entertaining, we may upset your sense of intuition.

Let us suppose you are in a room with about thirty-five people. What do you think the chances are (or probability is) of at least two of these people having the same birth date (month and day, only)? Intuitively, one usually begins to think about the likelihood of two people having the same date out of a selection of 365 days (assuming no leap year). Translating into mathematical language: 2 out of 365 would be a probability of $\frac{2}{365} = .005479 \approx \frac{1}{2}$ %. A minuscule chance.

197

Let's consider the "randomly" selected group of the first thirty-five presidents of the United States, since that number of people can represent the size of a rather large class of students. You may be astonished that there are two with the same birth date: the eleventh president, James K. Polk (November 2, 1795), and the twenty-ninth president, Warren G. Harding (November 2, 1865), shared the same birthday.

James K. Polk
Figure 5-1

Warren G. Harding
Figure 5-2

You may be surprised to learn that for a group of thirty-five, the probability that at least two members will have the same birth date is greater than 8 out of 10, or $\frac{8}{10} = 80\%$.

If you have the opportunity, you may wish to try your own experiment by selecting ten groups of about thirty-five people each to check on birthday matches. You ought to find that in about eight of these ten groups there will be a match of birth dates. For groups of thirty people, the probability that there will be a match is greater than 7 out of 10; or in seven of these ten rooms there would be a match of birth dates. What causes this incredible and unanticipated result? Can this be true? It seems to go against our intuition.

How can this probability be so high when there are 365 possible birthdates?[1] Let us consider the situation in detail and we will walk you through the reasoning that will eventually convince you that these are the true probabilities. Consider a class of thirty-five students. What do you think is the probability that one selected student matches his own birth date? Clearly *certainty*, or 1, since the first person has 365 possibilities to select his birth date out of 365 days.

This can be written as $\frac{365}{365}$

---

1. Actually 366, but for the sake of our illustration—and simplicity—we will not address the possibility of birthdays on February 29.

The probability that another student does *not* match the first student (i.e., has a different birth date) is $\frac{365-1}{365} = \frac{364}{365}$.

The probability that a third student does *not* match the first and second students is $\frac{365-2}{365} = \frac{363}{365}$.

The probability of all thirty-five students *not* having the same birth date is the *product* of these probabilities: $p = \frac{365}{365} \cdot \frac{365-1}{365} \cdot \frac{365-2}{365} \cdot \ldots \cdot \frac{365-34}{365}$.

Since the probability (q) that at least two students in the group *have* the same birth date and the probability (p) that *no* two students in the group have the same birth date is a certainty (i.e., there is no other possibility), the sum of those probabilities must be 1, which represents certainty.

Thus, $p + q = 1$, and it follows that $q = 1 - p$.

In this case, by substituting for p we get: $q = 1 - \frac{365}{365} \cdot \frac{365-1}{365} \cdot \frac{365-2}{365} \cdot \ldots \cdot \frac{365-34}{365}$ $\approx 0.8143832388747152$.

In other words, the probability that there will be a birth date match in a randomly selected group of thirty-five people is somewhat greater than $\frac{8}{10}$. This is, at first glance, quite unexpected when one considers there were 365 dates from which to choose. The motivated reader may want to investigate the nature of the probability function. Here are a few values to serve as a guide:

Number of people in group	Probability of a birth date match	Probability (in percent) of a birth date match
10	.1169481777110776	11.69 %
15	.2529013197636863	25.29 %
20	.4114383835805799	41.14 %
25	.5686997039694639	56.87 %
30	.7063162427192686	70.63 %
35	.8143832388747152	81.44 %
40	.891231809817949	89.12 %
45	.9409758994657749	94.10 %
50	.9703735795779884	97.04 %
55	.9862622888164461	98.63 %
60	.994122660865348	99.41 %
65	.9976831073124921	99.77 %
70	.9991595759651571	99.92 %

Notice how quickly almost-certainty is reached. With about sixty students in a room, the chart indicates that it is almost certain (99%) that two students will have the same birth date.

Were one to do this with the death dates of the first thirty-five presidents, one would notice that two died on March 8 (Millard Fillmore in 1874 and William H. Taft in 1930) and three presidents died on July 4 (John Adams and Thomas Jefferson in 1826 and James Monroe in 1831). Might this latter case be less a random event?

From the table above, we see that in a group of thirty people the probability of there being two people with the same birth date is about 70.63%. Yet, if we change this to the situation where you go into a room with thirty people, we can determine the probability of there being someone in the room with the same birthday as you to be about 7.9%—considerably lower, since we now seek a specific birth date, rather than just a match of any birth date.

Let's see how this can be found. We will determine the probability that there are no matches to your birth date and then subtract that probability from 1.

$$p \text{ Probability that no one has your birthday} = \left(\frac{364}{365}\right)^{30}$$

The probability that none of these thirty people has your birth date is then:

$$q = 1 - p \text{ Probability that no one has your birthday} = 1 - \left(\frac{364}{365}\right)^{30} \approx 0.079008598089550769$$

Perhaps even more amazing is that if you have a randomly selected group of two hundred people in a room, the probability of having two of these people born on the very same day (i.e., same year as well!) is about 50%.

Above all, this astonishing demonstration should serve as an eye-opener about the inadvisability of relying too much on intuition.

ANTICIPATING HEADS AND TAILS

This lovely little unit will show you how some clever reasoning along with algebraic knowledge of the most elementary kind will help you solve a seemingly "impossibly difficult" problem.

Consider the following problem.

You are seated at a table in a dark room. On the table there are twelve pennies, five of which are heads up and seven of which are tails up. (You know where the coins are, so you can move or flip any coin, but because it is dark you will not know if the coin you are touching was originally heads up or tails up.) You are to separate the coins into two piles (possibly flipping some of them) so that when the lights are turned on there will be an equal number of heads in each pile.

Your first reaction is "you must be kidding!" "How can anyone do this task without seeing which coins are heads or tails up? This is where a most clever (yet incredibly simple) use of algebra will be the key to the solution.

Let's "cut to the chase." (You might actually want to try it with 12 coins.) Separate the coins into two piles, of 5 and 7 coins each. Then flip over the coins in the smaller pile. Now both piles will have the same number of heads! That's all! You will think this is magic. How did this happen. Well, this is where algebra helps us understand what was actually done.

Let's say that when you separate the coins in the dark room, h heads will end up in the 7-coin pile. Then the other pile, the 5-coin pile, will have $5 - h$ heads. To get the number of tails in the 5-coin pile, we subtract the number of heads $(5 - h)$ from the total number of coins in the pile, 5, to get: $5 - (5 - h) = h$ tails.

5-coin pile	7-coin pile
$5 - h$ heads	h heads
$5 - (5 - h)$ tails $= h$ tails	

When you flip all the coins in the smaller pile (the 5-coin pile), the $(5 - h)$ heads become tails and the h tails become heads. Now each pile contains h heads!

The piles after flipping the coins in the smaller pile

5-coin pile	7-coin pile
$5 - h$ tails h heads	h heads

This absolutely surprising result will show you how the simplest algebra can explain a very complicated reasoning exercise.

PROBABILITY IN GEOMETRY

We hope not to upset your natural thought patterns with this example of what seems to be a very misleading probability. Consider two concentric circles where the radius of the smaller circle is one-half the radius of the larger circle (figure 5-3).

What is the probability that if a point *is selected in the larger circle* that it is also in the smaller one?

The typical (and correct) answer is $\frac{1}{4}$.

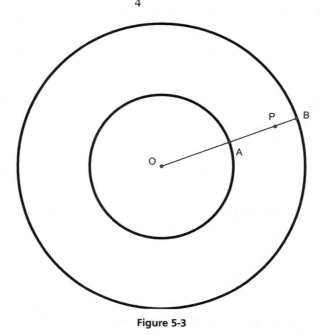

Figure 5-3

We know that the area of the smaller circle is $\frac{1}{4}$ the area of the larger circle:

$$r_{\text{smaller circle}} = \frac{1}{2} R_{\text{larger circle}} \text{ and } Area_{\text{larger circle}} = \pi (R_{\text{larger circle}})^2,$$

$$\text{and } Area_{\text{smaller circle}} = \pi (r_{\text{smaller circle}})^2 = \pi (\frac{1}{2} R_{\text{larger circle}})^2 = \frac{1}{4} \pi (R_{\text{larger circle}})^2$$

$$= \frac{1}{4} Area_{\text{larger circle}}$$

Therefore, if a point is selected at random in the larger circle, the probability that it would be in the smaller circle as well is $\frac{1}{4}$.

However, one might look at this question differently. The randomly selected point P must lie on some radius of the larger circle, say OAB, where A is its midpoint. The probability that a point P on OAB would be on OA (i.e., in the smaller circle) is $\frac{1}{2}$, since $OA = \frac{1}{2} OB$. Now, if we were to do this for any other point in the larger circle, we would find the probability of the point being in the smaller circle is $\frac{1}{2}$. This, of course, is *not* correct, although it seems perfectly logical. Where was the error in the second calculation made?

The "error" lies in the initial definition of each of two different sample spaces, that is, the set of possible outcomes of an experiment. In the first case, the sample space is the entire area of the larger circle, while in the second case, the sample space is the set of points on *OAB*. Clearly, when a point is selected on *OAB*, the probability that the point will be on *OA* is $\frac{1}{2}$. These are two entirely different problems even though (to dramatize the issue) they appear to be the same. The second "solution" is not representative of the problem, that a point is randomly selected from the *entire* circle.

FORMING A SQUARE

Our natural thinking often leads us to comfortable solutions. Then we hear the term "thinking out of the box" and we find that we must discard our normal thinking patterns. Here is a case when we will be thinking geometrically in a somewhat different way. This may surprise you at the end.

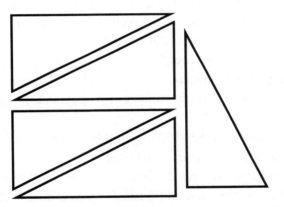

Suppose you have five congruent right triangles where one leg is twice the length of the other (figure 5-4). We must show how to form a square from these five right triangles, if you may only make one cut through one of the right triangles. It might be nice to try this without looking at the solution provided below.

Figure 5-4

However, trying to do this in the traditional way will just lead you to move the pieces around in a fruitless effort.

Yet, if we cut one of the right triangles along the perpendicular bisector of the longer leg (figure 5-5) and then place it as shown in figure 5-6, we will have formed a square as we were asked to do.

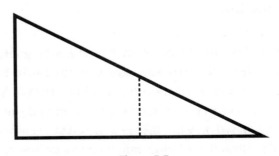

Figure 5-5

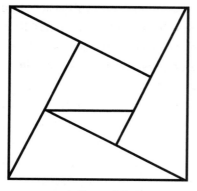

Figure 5-6

Notice how the two acute angles of the cut-up triangle are complementary (their sum is 90°), so they form a right angle. Also, the two angles formed by cutting the hypotenuse are supplementary (their sum is 180°), which is also required for our final square.

THE WORTHLESS INCREASE

Suppose you had a job where you received a 10% raise. Because business was falling off, the boss was soon forced to give you a 10% cut in salary. Will you be back to your starting salary? The answer is a resounding (and very surprising) NO!

This little story is quite disconcerting, since one would expect that with the same percent increase and decrease you should be where you started. This is intuitive thinking, but wrong. Convince yourself of this by choosing a specific amount of money and trying to follow the instructions.

Begin with $100. Calculate a 10% increase on the $100 to get $110. Now take a 10% decrease of this $110 to get $99—$1 less than the beginning amount.

You may wonder whether the result would have been different if we first calculated the 10% decrease and then the 10% increase. Using the same $100 basis, we first calculate a 10% decrease to get $90. Then the 10% increase yields $99, the same as before. So order appears to make no difference.

A gambler may face a similar situation that can be deceptively misleading. Consider the following situation, one that you may want to even simulate with a friend to see if your intuition bears out.

> **You are offered a chance to play a game. The rules are simple. There are 100 cards, face down. Of the cards, 55 say "win" and 45 say "lose." You begin with a bankroll of $10,000. You must bet one-half of your money on each card turned over, and you either win or lose that amount based on what the card says. At the end of the game, all cards have been turned over. How much money do you have at the end of the game?**

A similar principle as above applies here. It is obvious that you will win ten times more than you will lose, so it appears that you will end up with more than $10,000. What is obvious is not always right, and this is a good example. Let's say that you win on the first card. This means you win an additional half of your money, so you now have $15,000. Let's assume that you lose on the second card, which means that you lose half of the money you have so far, $15,000; so you now have $7,500. If you had first lost and then won, you would still have $7,500, since your money would go from $10,000 to $5,000 and then increase by $2,500. We can conclude from this that every time you win one and lose one, you will ultimately lose one-fourth of your money. This implies there are 45 win-lose times and then 10 extra win cards. You will, therefore, end up with $10,000 $\left(\frac{3}{4}\right)^{45} \left(\frac{3}{2}\right)^{10}$ = $10,000 \cdot (0.000137616\ldots)$. This is a mere $1.38 when rounded off. Surprised?

THE MONTY HALL PROBLEM ("LET'S MAKE A DEAL")

Let's Make a Deal was a long-running television game show that featured a problematic situation for a randomly selected audience member who was asked to come on stage and was presented with three doors. Two of the three doors had donkeys behind them, and one had a car, and the audience member was tasked with guessing which door concealed the car. There was only one wrinkle in this: after the contestant made a selection, the host, Monty Hall, knowing behind which door the car was, exposed one of the two donkeys behind a not-selected door (leaving two doors still unopened) and the audience participant was asked if she wanted to stay with her original selection (not yet revealed) or switch to the other unopened door. At this point, to heighten the suspense, the rest of the audience would shout out "stay" or "switch" with seemingly equal frequency. The question is what to do? Does it make a difference? If so, which is the better strategy (i.e., the greater probability of winning) to use here?

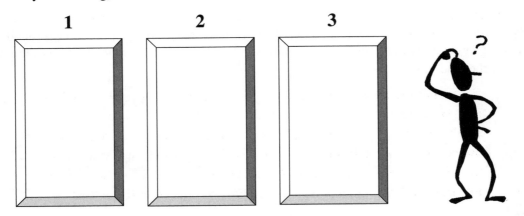

Figure 5-7

Let us look at this now step by step. The result may gradually become clear. There are two donkeys and one car behind these doors. You must try to get the car. You select Door #3. Monty Hall opens one of the doors that you *did not* select and exposes a donkey.

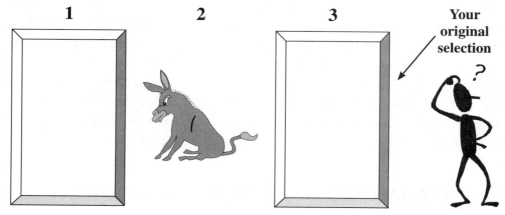

Figure 5-8

He asks: "Do you still want your first choice door, or do you want to switch to the other closed door"?

This is where the great confusion and controversy begins. To help make your decision, consider an *extreme case*:

Suppose there were one thousand doors instead of just three doors.

Figure 5-9

You choose Door #1,000. How likely is it that you chose the right door? ***Very unlikely,*** since the probability of getting the right door is $\dfrac{1}{1,000}$. How likely is it that the car is behind one of the other doors? ***Very likely***: $\dfrac{999}{1,000}$

Figure 5-10

These are all ***very likely*** doors!

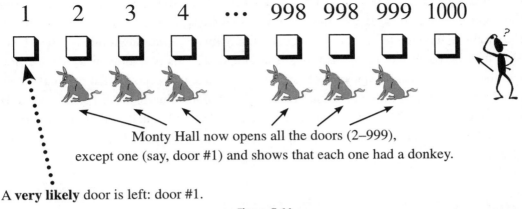

Monty Hall now opens all the doors (2–999),
except one (say, door #1) and shows that each one had a donkey.

A **very likely** door is left: door #1.

Figure 5-11

We are now ready to answer the question: Which is a better choice?:

♦ door #1,000 (*Very unlikely* door) or
♦ door #1 (*Very likely* door)?

The answer is now obvious. We ought to select the *very likely* door, which means "switching" is the better strategy for the audience contestant to follow.

Using the extreme case makes it much easier to see the best strategy, than had we tried to analyze the situation with the three doors. The principle is the same in either situation.

This problem has caused many an argument in academic circles and was also a topic of discussion in the *New York Times* and other popular publications. John Tierney wrote in the *New York Times* (Sunday, July 21, 1991) that "perhaps it was only an illusion, but for a moment here it seemed that an end might be in sight to the debate raging among mathematicians, readers of *Parade* magazine and fans of the television game show 'Let's Make a Deal.' They began arguing last September (Sept. 9, 1990) after Marilyn vos Savant published a puzzle in *Parade*. As readers of her 'Ask Marilyn' column are reminded each week, Ms. vos Savant is listed in the Guinness Book of World Records Hall of Fame for 'Highest I.Q.,' but that credential did not impress the public when she answered this question from a reader." She gave the right answer, but still many mathematicians argued. Even Paul Erdös (1913–1996), one of the most prolific mathematicians of the twentieth century, was stumped by this problem and had to be convinced about the error of his thinking.[2]

You might want to see if you have been properly alerted to this situation with the following problem. Suppose you have three cards: one has two blue sides, one has two red sides, and the third card has one red side and one blue side. Without looking at the bottom side of one of the three cards, one is placed on a table. If the side face up (showing) is blue,

---

2. Paul Hoffman, *The Man Who Loved Only Numbers* (New York: Hyperion, 1998), pp. 249–56.

then clearly this card is not the red-red card. If it is the blue-blue card, then the bottom (face-down) side is blue. If the card is the blue-red card, then the face-down side is red. You might be tempted to say that it is equally likely that the face-down side is red or blue. Well, by now you might realize that this is not the case. This is analogous to the decision with the two doors earlier. The correct analysis must take into account that we might have been looking at side 1 of the blue-blue card or side 2 of the blue-blue card or the blue side of the blue-red card. There are three possible blue sides—all equally likely. Two of them will lead to a face-down blue side and one to a face-down red side. Therefore, red or blue are not equally likely.

SHARPENING PROBABILITY THINKING

Coin tossing case 1:

Suppose we toss four coins, what is the probability that there will be at least two heads? A description of the possible outcomes will bring some clarity to the question (H = heads; T = tails).

Number of heads	Possible outcomes by order of toss
4	HHHH
3	HHHT, HHTH, HTHH, THHH
2	HHTT, HTHT, TTHH, THHT, HTTH, THTH
1	TTTH, TTHT, THTT, HTTT
0	TTTT

There were eleven (i.e., $1 + 4 + 6 = 11$) tosses that resulted in at least two heads out of sixteen possible resulting tosses. This means that the probability of getting at least two heads is $\frac{11}{16}$ = 0.6875.

Coin tossing case 2:

This time we will toss a coin ten times. The first time resulted in the following:

    1.   H, T, T, H, T, T, T, H, T, H

The second time we tossed the coin ten times we got the following:

    2.   H, T, H, T, H, T, H, T, H, T

The third time we tossed the ten coins our result was:

    3.   H, H, H, H, H, H, H, H, H, H

The question we are faced with is: Which of the three tosses is most likely to occur? One is tempted to select outcome no. 1, since it is assumed that unordered results are most likely to occur. This assumption is false! A nicely ordered result (such as nos. 2 and 3) is just as likely to occur as one that is unordered—as no. 1 is.

Since, for each toss, the probability of a head and a tail is equally likely—namely, a probability of $\frac{1}{2}$, the probability of each of the three results is equally likely with a probability of $\left(\frac{1}{2}\right)^{10} = \frac{1}{1,024} = 0.0009765625$.

The famous French mathematician Jean le Rond d'Alembert (1717–1783) falsely assumed that "if you have already tossed nine heads then it is more likely that a tail will be tossed on the tenth try." Yet this assumption was challenged by another French mathematician, Pierre Raymond de Montmort (1678–1719), when he said that "the past does not determine the future." Further chiming in to the debate was the famous Swiss mathematician Leonhard Euler (1707–1783), who said, "Then every toss would be dependent on every previous toss, regardless in which town it occurred, even if it happened one hundred years earlier—which is the most absurd thinking imaginable."

We must take into account that in the first case order of the result was not important, whereas in the second case it was. These are some of the subtle differences that must be accounted for when we look at questions of probability.

INTRODUCTION OF A SAMPLE SPACE

We have a tendency, when faced with a mathematical problem (or game), to search for an automatic way to solve the problem. This can, on occasion, be counterproductive, or just not helpful. When doing a problem involving probability, it can be wise to set up the sample space to see what is actually taking place. This activity could place you in a game situation, where your intuition could work against you. Unless you actually set up the sample space, you may not be able to resolve the inequity of the game we are about to present to you.

Begin by placing 1 red chip and 2 black chips in an envelope. Here are the rules for the game we are about to play with the reader against the authors:

1. You draw 2 chips from the envelope, without looking.
2. If the colors of the 2 chips are different, we score a point. If they are the same, you score the point.
3. The first player to score 5 points is the winner.

4. After each draw, the chips are returned to the envelope and the envelope is shaken.

Do you believe that the game is a fair one (i.e., each player has an equal chance of gaining a point)? You might actually try this game with a friend—who can take the place of the authors. After playing the game several times, you might conclude that the game is not fair. The authors (or their surrogate) should win the game most of the time. What single chip would you add to make the game fair? Typically, one would suggest adding a second red chip to the envelope. It might be best to look at this problem with the help of setting up a sample space.

Situation 1: 1 red chip and 2 black chips
The possible draws would be:

$$RB_1 \qquad RB_2 \qquad \mathbf{B_1B_2}$$

Thus, you have only 1 out of 3 chances of scoring a point, for a $\frac{1}{3}$ probability. Therefore, the original game is unfair.

Situation 2: Adding 1 red chip, we have 2 red chips and 2 black chips
The possible draws would be:

$$R_1B_1 \qquad R_1B_2 \qquad \mathbf{R_1R_2}$$
$$R_2B_1 \qquad R_2B_2 \qquad \mathbf{B_1B_2}$$

Surprise! You have only 2 out of 6 chances of scoring a point, for a $\frac{1}{3}$ probability. The game is, once again, unfair.

Situation 3: Adding another black chip to the original chips, we get 1 red chip and 3 black chips
The possible draws would be:

$$R_1B_1 \qquad R_1B_2 \qquad R_1B_3$$
$$\mathbf{B_1B_2} \qquad \mathbf{B_1B_3} \qquad \mathbf{B_2B_3}$$

This time you have 3 out of 6 chances of scoring a point, for a $\frac{1}{2}$ probability. The game is now fair.

The use of the sample space easily reveals that your intuition does not yield a correct resolution of the problem, thus making the concept of a sample space "indispensable."

USING SAMPLE SPACES TO SOLVE TRICKY PROBABILITY PROBLEMS

Now that you can see how identifying a sample space can be useful, we shall try to further convince you of its value by showing you another problem that would defy intuition. Consider the following problem.

> **A person has just three phonograph records. The first has vocal performances on both sides, the second has instrumental music on both sides, and the third has vocal performances on one side and instrumental music on the other side. This person, who is in a darkened room, plays one of these records. What is the probability that he hears a vocal performance?**

To make this problem a bit more manageable, we shall use some symbols and

$$\text{denote by } v_{1,1}: \text{side 1 of record 1}$$
$$\text{denote by } v_{2,1}: \text{side 2 of record 1}$$
$$\text{denote by } i_{1,2}: \text{side 1 of record 2}$$
$$\text{denote by } i_{2,2}: \text{side 2 of record 2}$$
$$\text{denote by } v_{1,3}: \text{side 1 of record 3}$$
$$\text{denote by } i_{2,3}: \text{side 2 of record 3}$$

The sample space for this problem situation consists of the six equally likely possible outcomes: $v_{1,1}$, $v_{2,1}$, $i_{1,2}$, $i_{2,2}$, $v_{1,3}$, and $i_{2,3}$. Precisely three of which: $v_{1,1}$, $v_{2,1}$, and $v_{1,3}$, consist of vocal performances. Therefore, the probability of hearing a vocal performance is $\frac{3}{6} = \frac{1}{2}$. This is a reasonable solution—one that conforms with our intuition.

Now let's consider a somewhat more difficult problem.

> **A person has just three phonograph records. The first has vocal performances on both sides, the second has instrumental music on both sides, and the third has vocal performances on one side and instrumental music on the other side. This person, who is in a darkened room, puts on one of these records and hears a vocal performance. What is the probability that the other side of that same record is also vocal?**

The usual response is $\frac{1}{2}$, with the "reasoning" that one of the two vocal-sided records has a vocal on the other side. This reasoning is wrong!

The correct reasoning is that since there are six record sides equally likely to be played and the person in the problem is playing a vocal side, the sample space for this is limited to the set $\{v_{1,1}, v_{2,1}, v_{1,3}\}$. Two of the three elements, $\{v_{1,1}, v_{2,1}\}$, of the sample space set represent a vocal on the other side. Therefore, the probability is $\frac{2}{3}$. This easily missolved problem demonstrates the usefulness of setting up a sample space. Compare this "conditional probability" $\left(\frac{2}{3}\right)$ with the probability $\left(\frac{1}{2}\right)$ of the first problem above.

PROBABILITY THROUGH COUNTING

There is a thinking adjustment required when one first encounters probability. A simple illustration, such as pulling an ace out of a deck of fifty-two cards, requires little imagination. However, a problem such as the following does require some probability-thinking skills.

Consider the following problem:

> **Two identical paper bags contain red and black checkers.**
> > **<u>Bag A</u> contains 2 red and 3 black checkers.**
> > **<u>Bag B</u> contains 3 red and 4 black checkers.**
> **A bag is chosen at random and a red checker is drawn from it. What is the probability that Bag A was chosen?**

This figure (5-12) should help your thought process as you contemplate the problem.

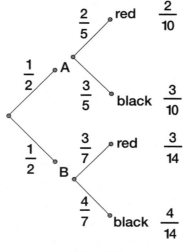

Figure 5-12

Since the first even common multiple of 5 and 7 (the bags' checker contents) is 70, we shall use this as the number of trials for our hypothetical experiment. Since the bag was chosen at random, we can assume for our convenience that each bag was chosen 35 times.

In that case:

From <u>Bag A</u>: a red checker will be selected 2 out of 5 times.
     So for 35 trials a red checker will be drawn 14 times.

From <u>Bag B</u>: a red checker will be selected 3 out of 7 times.
     So for 35 trials a red checker will be drawn 15 times.

Therefore, a red checker will be drawn $14 + 15 = 29$ times, of which 14 would be from Bag A. Consequently, the probability that the red checker will be drawn from Bag A is $\frac{14}{29}$.

This displays the kind of comparison useful to succeed in determining the probability of an event.

SENSIBLE COUNTING

The technique of systematic counting can be very useful. It can also lead us to some unexpected results. Consider the task of marking each of the faces of a cube with 1, 2, 3, 4, 5, and 6 dots, with the stipulation that the 1 and the 6 must be on opposite faces, the 2 and the 5 opposite each other, and the 3 and 4 on opposite faces of the cube as well. This seems easy enough but can be somewhat perplexing, if you do not approach the problem in a systematic fashion. You will be surprised at the many ways this task can be accomplished.

Suppose we mark any one face with the 1 dot. This, of course, can be done in any one of <u>6</u> ways. Then we know that the 6 must be opposite it. This leaves <u>4</u> possible faces to place the 2 dots. Once the 2 has been placed, the 5 has also been established, since it is to be on the opposite face. This, then, leaves us with <u>2</u> possible placements for the 3. To find the total number of ways to mark the cube with dots—as previously required—we simply multiply the three placement possibilities: $6 \cdot 4 \cdot 2 = 48$.

Such organized counting can be helpful in determining ways in which things can be accomplished, and oftentimes they are counterintuitive.

ORGANIZED COUNTING

Organized counting can also be useful in a geometrical setting. Suppose you are asked to tear off four postage stamps from a sheet of twelve stamps arranged in three rows of four stamps each. The four stamps that you tear off must be attached on at least one side to the others. In how many ways can you tear off these four stamps? This problem will take you through a nifty method of organized counting. Consider the sheet of stamps as in figure 5-13.

A	B	C	D
E	F	G	H
J	K	L	M

Figure 5-13

What we need to focus on is the various shapes that the set of four stamps can have, and then we can count the number of ways that shape can be found on the sheet:

Type ABCD: 3

Type ABFE: 6

Type EABC, ABCG, AEFG, EFGC, ABEJ, AEJK, ABFK, or BFKJ: 28

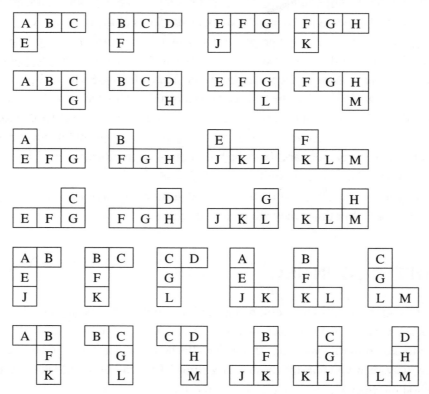

Type ABCF, BEFG, AEFJ, or BEFK: 14

Type ABFG, CBFE, AEFK, or BFEJ: 14

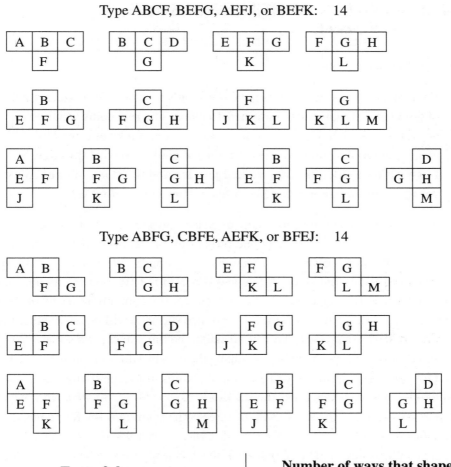

Type of shape	Number of ways that shape can be formed
ABCD	3
ABFE	6
EABC, ABCG, AEFG, EFGC, ABEJ, AEJK, ABFK, or BFKJ	28
ABCF, BEFG, AEFJ, or BEFK	14
ABFG, CBFE, AEFK, or BFEJ	14
TOTAL:	**65**

Figure 5-14

This sort of systematic counting is useful in getting us organized and lets us appreciate a rather surprising result.

HOW ORGANIZED COUNTING CAN SOLVE A BAFFLING PUZZLE

Consider the following problem:

> **Each of the 10 court jewelers gave the king's advisor, Mr. Loge, a stack of gold coins. Each stack contained 10 coins. The real coins weighed exactly 1 ounce each. However, one, and only one, stack contained "light" coins, each having had exactly 0.1 ounce of gold shaved off the edge. Mr. Loge wishes to identify the crooked jeweler and the stack of light coins with just one single weighing on a scale. How can he do this?**

The traditional procedure is to begin by selecting one of the stacks at random and weighing it. This trial-and-error technique offers only a 1 chance in 10 of being correct. You may then attempt to solve the problem by reasoning: if all the coins were true, their total weight would be (10)(10), or 100 ounces. Each of the 10 counterfeit coins is lighter, so there will be a deficiency of (10)(0.1), or 1 ounce. But thinking in terms of the overall deficiency doesn't lead anywhere, since the 1 ounce shortage will occur whether the counterfeit coins are in the first stack, the second stack, the third stack, and so on.

Let us try to solve the problem by organizing the data in a different fashion. We must find a method for varying the deficiency in a way that permits us to identify the stack from which the counterfeit coins are taken. Label the stacks #1, #2, #3, #4, . . . #9, #10. Now, we take one coin from stack #1, two coins from stack #2, three coins from stack #3, four coins from stack #4, and so on. We now have a total of $1 + 2 + 3 + 4 + . . . + 8 + 9 + 10 = 55$ coins. If they were all true, the total weight would be 55 ounces. If the deficiency were .5 ounces, then there were 5 light coins, taken from stack #5. If the deficiency were .7 ounces, then there were 7 light coins, taken from stack #7, and so on. Thus, Mr. Loge could readily identify the stack of light coins, and consequently determine which jeweler shaved each coin.

ORGANIZED COUNTING CAN BRING ORDER WHEN IT IS ESSENTIAL

How many triangles are in figure 5-15?

The traditional method—or that which "might be expected by the teacher"—would depend on formal counting methods. These involve calculating the combinations that can be formed by the six lines, excluding those combinations resulting in concurrency. Hence, the number of combinations of six lines, taken three at a time, yields $_6C_3 = 20$ combinations. From this we subtract the 3 concurrencies (at the vertices). Thus, there are 17 triangles in figure 5-15.

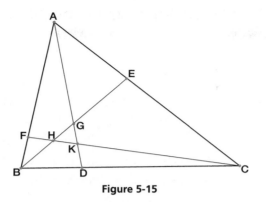

Figure 5-15

If students attempt to count the triangles in the figure, they could very likely miss some of them in their counting. It is obvious that they need some method of organizing the information to obtain an accurate answer.

Let's try to simplify the problem by reconstructing the figure, gradually adding the lines as we go, and counting from this form of *organized data*—that is, counting the triangles created by the addition of each additional part of the figure.

Start with the original triangle, $\triangle ABC$ (figure 5-16). Thus, we have exactly 1 triangle.

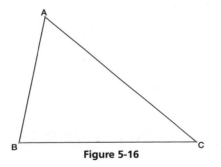

Figure 5-16

Now consider $\triangle ABC$ with one interior line, AD (figure 5-17). We now have 2 *new* triangles, $\triangle ABD$ and $\triangle ACD$.

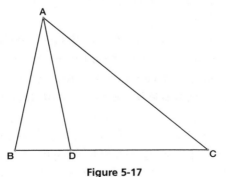

Figure 5-17

Now add the next interior line, *BE* (figure 5-18), and count all the *new* triangles that utilize *BE* as a side: $\triangle ABG$, $\triangle BDG$, $\triangle AEG$, $\triangle BCE$, and $\triangle ABE$.

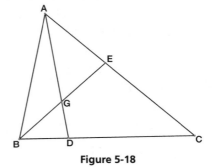

Figure 5-18

Continue in this manner, adding line *CF* (figure 5-19). Again count the new triangles using part of *CF* as a side: $\triangle BFH$, $\triangle ACF$, $\triangle BCH$, $\triangle AFK$, $\triangle CDK$, $\triangle ACK$, $\triangle BCF$, $\triangle GHK$, and $\triangle CEH$.

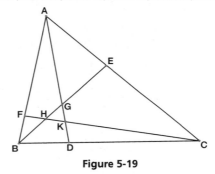

Figure 5-19

There are seventeen triangles in the figure.

UNDERSTANDING THE CONCEPT OF RELATIVITY

Although sometimes difficult to grasp, the concept of relativity is one that we often encounter in everyday life and do not properly comprehend. Let's consider the following question and see if it helps us appreciate this idea.

> **While rowing his boat upstream, David drops a basketball overboard and continues rowing for ten more minutes. He then turns around, chases the basketball, and retrieves it when the basketball has traveled one mile downstream. What is the rate of the stream?**

Rather than approach this problem through traditional methods common in an algebra course, we shall consider the following method. We have here an example of the notion of relativity. It does not matter if the stream is moving and carrying David downstream or is still. We are concerned only with the separation and coming together of David and the basketball. If the stream were stationary, David would require as much time rowing to the basketball as he did rowing away from the basketball. That is, he would require 10 + 10 = 20 minutes. Since the basketball travels one mile during these 20 minutes, the stream's rate of speed is 3 miles per hour.

This may not be an easy concept for some to grasp; yet ponder it in quiet, since it is a concept worth understanding, and it has many useful applications in everyday life thinking processes. This is, after all, one of the many purposes for learning mathematics.

WHERE IN THE WORLD ARE YOU?

There are entertainments in mathematics that stretch the mind (gently, of course) in a very pleasant and satisfying way. Such examples can leave us in amazement, which further generates our appreciation of mathematics. We present just such a situation now. This popular puzzle question, which has some very interesting extensions, requires some "out of the box" thinking.

Where on earth can you be so that you can walk *one mile south*, then *one mile east*, and then *one mile north* and end up at the starting point? (See figure 5-20.)

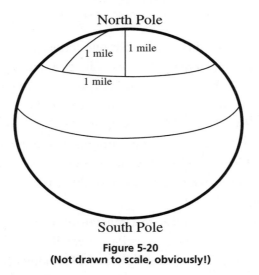

Figure 5-20
(Not drawn to scale, obviously!)

Most people will approach this problem with a trial-and-error method. We take you to the correct answer immediately to avoid frustration: the North Pole. To test this answer, try starting from the North Pole and traveling south one mile. Then, traveling east one mile, takes you along a latitudinal line that remains equidistant from the North Pole, one mile from it. Traveling one mile north gets you back to where you began, the North Pole.

Most people familiar with this problem feel a sense of satisfaction. Yet we can ask: Are there other such starting points where we can take the three same-length "walks" and end up at the starting point? The answer, surprisingly enough for most people, is *yes*.

One set of starting points is found by locating the latitudinal circle, which has a circumference of one mile and is nearest the South Pole. From this circle, walk one mile north (along a great circle route, naturally),[3] and form another latitudinal circle. Any point along this second latitudinal circle will qualify. Let's try it (figure 5-21).

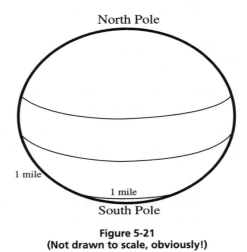

North Pole

1 mile

1 mile

South Pole

Figure 5-21
(Not drawn to scale, obviously!)

Begin on this second latitudinal circle (the one farther north). Walk one mile south (takes you to the first latitudinal circle), then one mile east (takes you exactly once around the circle), and then one mile north (takes you back to the starting point).

Suppose the second latitudinal circle, the one we would walk along, would have a circumference of $\frac{1}{2}$ mile. We could still satisfy the given instructions, by this time walking around the circle *twice*, and end up back at our original starting point. If the second latitudinal circle had a circumference of $\frac{1}{4}$ mile, then we would merely have to walk around this circle *four* times to get back to the starting point of this circle and then go north one mile to the original starting point.

At this point we can take a giant leap to a generalization that will lead us to many more points that would satisfy the original stipulations. Actually, an infinite number of points! This set of points can be located by beginning with the latitudinal circle, located nearest the South Pole, which has a $\frac{1}{n}$ th -mile circumference. An n-mile walk east will take you back to the point on the circle at which you began your walk on this latitudinal circle. The rest is the same as before, that is, walking one mile south and then later one mile north.

---

3. The great circle of a sphere is the largest circle that can be drawn on the surface of a sphere. It is one formed by a plane intersecting the sphere and containing the center of the sphere.

HOW MANY TIMES CAN YOU FOLD A PIECE OF PAPER?

The question about how many times one can fold a piece of paper with the condition that the paper gets folded in half (through its center) continuously presents a physical amazement that can be best comprehended arithmetically. We can add another question onto this first question, namely, how many times must you fold a piece of paper (with each new fold through the remaining center) so that the resulting thickness will equal the distance from Earth to the Moon? Of course, this last question is mostly theoretical, since it could not be physically carried out.

Try taking a piece of paper of normal thickness and see how many times you can fold it with each succeeding fold dividing the remaining rectangle in half. You will find (surprisingly) that the maximum number of folds will be eight.

(This results far less from the paper thickness than from the friction between the folded surfaces and the friction between the paper and your fingers.)

Now, suppose that theoretically we were able to continue folding a piece of paper, which has a thickness of 0.1 mm, forty-two times. After each fold, the paper would double its thickness. After eight folds, the paper's thickness would be 0.1 mm $\cdot 2 \cdot 2 \cdot 2 \cdot 2 \cdot 2 \cdot 2 \cdot 2 \cdot 2$ = 0.1 mm $\cdot 2^8$ = 0.1 mm $\cdot 256$ = 25.6 mm, or 2.56 cm.

With the ninth fold, you would be required to fold a piece of paper with the thickness of 2.56 cm, which is impossible with the typical finger strength. After the (theoretical) forty-two folds, we would gain a thickness of 0.1 mm $\cdot 2^{42}$ = 0.1 mm $\cdot 4,398,046,511,104$ = 43,980,465,111.04 cm = 439,804,651.1104 m $\approx$ 439,804.651 km. This exceeds the distance of the Moon from Earth, which is approximately 384,400 km.

HOW CAN 64 = 65?

Consider an 8 × 8 square (perhaps, using convenient graph paper). Then cut up the square as in figure 5-22. Next place the pieces together as shown in figure 5-23. We have formed a rectangle of dimensions 5 × 13.

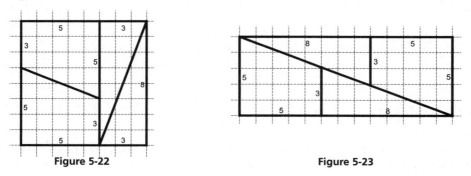

Figure 5-22   Figure 5-23

Since we used the same pieces, we would likely assume that the two rectangles in figures 5-22 and 5-23 must have the same area. Yet, by calculating the area of each of these figures, we find that the square has an area of $8 \cdot 8 = 64$ and the rectangle has an area of $5 \cdot 13 = 65$. How can this be? They should have the same area. How did we lose a unit square when we used the same pieces to make each of the figures?

The answer lies in the fact that the pieces we used to create the rectangle in figure 5-23 do not really fit together as shown—it is an optical disappointment!

A more accurate picture of the attempted construction of the rectangle reveals that the angles that we just assumed fit together to form right angles truly do not (see figure 5-24).

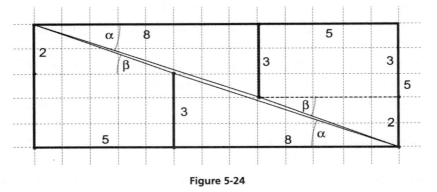

Figure 5-24

If we use the tangent trigonometric ratio, we can get the angle measure very easily.

The $\tan \alpha = \frac{3}{8}$, or $\alpha \approx 20.6°$; the $\tan \beta = \frac{2}{5}$, or $\beta \approx 21.8°$; that is, the difference between the two angles, β and α, which should be equal (in order for us to have a true rectangle), is rather small: $\beta - \alpha \approx 1.2°$.

This is an angle that is not easily discernible with normal eyesight.

An informed reader may recognize the appearance of the Fibonacci numbers: 1, 1, 2, 3, 5, 8, 13, ...[4]

SOLVED AND UNSOLVED PROBLEMS

Who says that all mathematical problems get solved? Unsolved problems have a very important role in mathematics. Attempts to solve them sometimes lead to significant findings of other sorts. An unsolved problem, one not yet solved by the world's most brilliant minds, tends to pique our curiosity by quietly asking us if we can solve it, especially when the problem itself seems exceedingly easy to understand.

4. For more on these ubiquitous numbers see A. S. Posamentier and I. Lehmann, *The Fabulous Fibonacci Numbers* (Amherst NY: Prometheus Books, 2007), in particular pp. 142–43.

The German mathematician David Hilbert (1862–1943) announced his seminal list of twenty-three unsolved mathematical problems at the second International Congress of Mathematicians in Paris on August 8, 1900. Much of the mathematical research in the twentieth century was influenced by this list of unsolved problems, and both successful and unsuccessful attempts at solutions yielded a number of important discoveries along the way.

To commemorate this memorable occasion and to provide a suitable launch for mathematics into the twenty-first century (one hundred years later), the newly formed Clay Mathematics Institute, located in Cambridge, Massachusetts, then devised its own list of seven problems that still lacked a solution. It made its announcement at the Collège de France in Paris on May 24, 2000, in a lecture titled "The Importance of Mathematics." The founder and sponsor of the Clay Mathematics Institute, Landon T. Clay, a businessman who majored in mathematics at Harvard University, believes mathematics research is underfunded and is dedicated to popularizing the subject. He offered $1,000,000 to anyone who can solve any of the listed heretofore-unsolved mathematics problems. In 2002, one of these seven problems, the Poincaré Conjecture, was solved by the Russian mathematician Grigori Perelman (1966–), who refused the prize. He was also awarded the Fields Medal[5] in 2006 and also declined to accept it.

We shall look at some unsolved problems to get a better understanding of the history of mathematics. Four times, in recent years, mathematics has made newspaper headlines —each time with the solution to a longtime unsolved problem.

The *Four-Color Map Problem* dates back to 1852 when Francis Guthrie (1831–1899), while trying to color the map of counties of England, noticed that four colors sufficed. He asked his brother Frederick if it was true that **any** map can be colored using only four colors in such a way that adjacent regions (i.e., those sharing a common boundary segment, not just a point) receive different colors. Frederick Guthrie then communicated the conjecture to the famous mathematician Augustus de Morgan (1806–1871).

As early as 1879, the British mathematician Alfred B. Kempe (1849–1922) produced an attempted proof, but in 1890 it was shown to be wrong by Percy J. Heawood (1861–1955). Many other subsequent attempts also proved fallacious.

Not until 1976 was this so-called four-color map problem solved by two mathematicians, Kenneth Appel (1932–) and Wolfgang Haken (1928–), who, using a computer, considered all possible maps and established that it was never necessary to use more than four colors to color a map so that no two territories, sharing a common border, would be represented by the same color.[6] They used an IBM 360 that required about 1,200 hours to test the 1,936 cases, which later turned out to require inspecting another 1,476 cases.

---

5. The Fields Medal is the highest prize in mathematics awarded to mathematicians under forty years of age.
6. Kenneth Appel and Wolfgang Haken, "The Solution of the Four-Color-Map Problem," *Scientific American* 237, no. 4 (1977): 108–21.

This "computer proof" was not widely accepted by pure mathematicians. Yet, in 2004, the mathematicians Benjamin Werner and Georges Gonthier produced a formal mathematical solution to the four-color map problem that validated Appel and Hacken's assertion.

The attractive aspect of the four-color map problem lies in the fact that it is very easily understood, but the solution has proved to be very elusive and very complex.

Then there is ***Fermat's Last Theorem.*** More recently, on June 23, 1993, Andrew Wiles (1953–), a Princeton University mathematics professor, announced that he solved the 350-year-old "Fermat's Last Theorem." It took him another year to fix some gaps in the proof, but it put to rest a nagging problem that had occupied scores of mathematicians for centuries. The statement of a "theorem," which Pierre de Fermat (1607–1665) wrote (ca. 1630) in the margin of a mathematics book he was reading (Diophantus' *Arithmetica*), was discovered by his son after his death. In addition to this statement, Fermat indicated that his proof was too long to fit in the margin of the book, so he effectively left to others the job of proving his statement. To this day, we wonder if Fermat really did have a proof, or if this was put there as a joke for posterity. Some say that he may have thought he had a proof, but it might not have been correct. In any case, let's take a look at what this famous theorem stated.

Fermat's "Theorem"[7]: $x^n + y^n = z^n$ **has no nonzero integer solutions for** $n > 2$.

Fermat did show that for the value on $n = 4$, his conjecture (or supposed "theorem") was correct. Leonhard Euler (1707–1783), the prolific Swiss mathematician, took this one step further by showing that "Fermat" was correct for the case when $n = 3$. In 1825, Peter Gustav Lejeune-Dirichlet (1805–1859) and Adrien-Marie Legendre (1752–1833) extended the conjecture a step further to the case where $n = 5$. Then in 1839 Gabriel Lamé (1795–1870) showed that "Fermat" was also correct for $n = 7$. The number of cases for which Fermat's conjecture was shown to hold true kept increasing: in 1857 Ernst Eduard Kummer (1810–1893) showed it true for all cases of $n \leq 100$. Harry Vandiver (1882–1973) took this even further in 1937 to all cases of $n \leq 617$. With the aid of a computer, it progressed even further: in 1954 for $n \leq 2,500$; in 1976 for $n \leq 125,000$; in 1987 for $n \leq 150,000$; and in 1991 for $n \leq 1,000,000$.

Then, finally, in 1993, during a multiday lecture, Andrew Wiles surprised a group of mathematicians with a proof of the general case of the conjecture. That is, he showed that Fermat's statement was true for all values of n and hence it is then a true theorem.

---

7. In other words, there are no integer values for x, y, and z for which this equation will hold true, with the exception when $n \leq 2$.

The third time that a previously unsolved problem had been solved and made newspaper headlines was in 1998 when the American mathematician Thomas C. Hales (1958–) proved the four-hundred-year-old *Kepler Conjecture*[8] using a computer.

During this time speculation began about other unsolved problems of which many still exist. Two of them are very easy to understand but are apparently exceedingly difficult to prove. Neither has yet been proved. We shall present some others here. Remember, although the best mathematicians haven't been able to prove these relationships, we still have never found counterexamples to allow us to say that they are not always true. Computers have been able to generate enormous numbers of examples to support the following statement's veracity, but that (amazingly) does not make it true for all cases. To conclude the truth of a statement, we must have a legitimate proof that it is true for all cases!

Christian Goldbach (1690–1764), a German mathematician, in a June 7, 1742, letter to Leonhard Euler (1707–1783), posed the following statement, which to this day has yet to be solved. *Goldbach's Conjecture* is as follows:

**Every even number greater than 2 can be expressed as the sum
of two prime numbers.**

You might want to begin with the following list of even numbers and their prime number sums and then continue it to convince yourself that it continues on—apparently—indefinitely (figure 5-25).

Even numbers greater than 2	Sum of two prime numbers
4	2 + 2
6	3 + 3
8	3 + 5
10	3 + 7
12	5 + 7
14	7 + 7
16	5 + 11
18	7 + 11
20	7 + 13
…	…
48	19 + 29
…	…
100	3 + 97

Figure 5-25

8. The Kepler Conjecture stated that no arrangement of equally sized spheres filling space has a greater average density than the cubic close packing and the hexagonal close packing arrangements.

Again, there have been substantial attempts by famous mathematicians: In 1855, A. Desboves verified Goldbach's conjecture for up to 10,000 places. Yet in 1894, the famous German mathematician Georg Cantor (1845–1918) (regressing a bit) showed that the conjecture was true for all even numbers up to 1,000; it was then shown by N. Pipping in 1940 to be true for all even numbers up to 100,000. By 1964, with the aid of a computer, it was extended to 33,000,000; in 1965 to 100,000,000; and then in 1980 to 200,000,000. In 1998, the German mathematician Jörg Richstein showed that Goldbach's Conjecture was true for all even numbers up to 400 trillion. On February 16, 2008, Oliveira e Silva extended this to 1.1 quintillion (i.e., $1.1 \times 10^{18} = 1,100,000,000,000,000,000$)! Prize money of $1,000,000 has been offered for a proof of this conjecture. To date, this has not been claimed.

Goldbach's *Second Conjecture* is as follows:

Every odd number greater than 5 can be expressed as the sum of three primes.

Again, we shall present you with a few examples and let you continue the list as you wish (figure 5-26).

Odd numbers greater than 5	Sum of three prime numbers
7	2 + 2 + 3
9	3 + 3 + 3
11	3 + 3 + 5
13	3 + 5 + 5
15	5 + 5 + 5
17	5 + 5 + 7
19	5 + 7 + 7
21	7 + 7 + 7
...	...
51	3 + 17 + 31
...	...
77	5 + 5 + 67
...	...
101	5 + 7 + 89

Figure 5-26

These unsolved problems have tantalized many mathematicians over the centuries, and although no proof has yet been found, more evidence (built with the help of computers) suggests that these must be true, since no counterexamples have been found. Interestingly, the efforts to solve them have led to some significant discoveries in mathematics that might have gone hidden without this impetus.

They provoke us and provide sources of entertainment.

Another perplexing question in mathematics involves *the edges and diagonals of a rectangular solid*. The question is: Can a rectangular solid be constructed where the three edges (length, width, and height) have integer lengths and the distance between any two vertices is also an integer length? (See figure 5-27.)

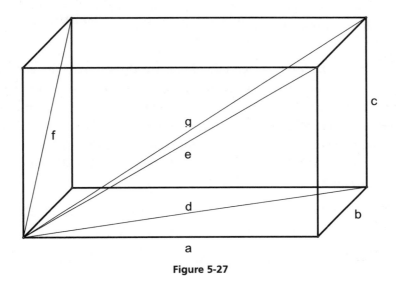

Figure 5-27

To check on this, we must be able to apply the Pythagorean Theorem to show that each of the following equations can be solved with only integers:

$$a^2 + b^2 = d^2$$
$$a^2 + c^2 = e^2$$
$$b^2 + c^2 = f^2$$
$$a^2 + b^2 + c^2 = g^2$$

Leonhard Euler almost found a solution to this problem with the following values:
$a = 240, b = 44, c = 117, d = 244, e = 267, f = 125$.

However, this fell short of the complete solution to the problem, since the last of these equations ($a^2 + b^2 + c^2 = g^2$) does not result in an integer solution; $g \approx 270.60$. It was not until the year 2000 that Marcel Lüthi, a Swiss mathematics teacher, proved that such a rectangular solid with integer distances between all vertices could not exist.[9] Again, we have a situation where a problem remained unsolved for several hundred years.

---

9. Marcel Lüthi, "Zum Problem rationaler Quader," *Praxis der Mathematik* 42, no. 4 supplement (2000): 177.

LOGICAL THINKING

When a problem is posed that at first looks a bit daunting and then a solution is presented —one easily understood—we often wonder why we didn't think of that simple solution ourselves. Such problems that have a gee-whiz dramatic effect on us will likely help us with future (analogous) situations. Here is one such problem.

> **On a shelf in Danny's basement, there are three jars. One contains only nickels, one contains only dimes, and one contains a mixture of nickels and dimes. The three labels, "nickels," "dimes," and "mixed," fell off, and were all put back on the wrong jars (see figure 5-28). Without looking, Danny can select one coin from one of the mislabeled jars and then correctly label all three jars. From which jar should Danny select the coin?**

Nickels   **Dimes**   **Mixed**

Figure 5-28

One may reason that the "symmetry" of the problem dictates that whatever we can say about the jar mislabeled "nickels" could just as well have been said about the jar mislabeled "dimes." Thus, if Danny chooses a coin from either of these jars, the results would be the same.

You should, therefore, concentrate your considerations on what happens if he chooses from the jar mislabeled "mixed." Suppose Danny selects a nickel from the "mixed" jar. Since this jar is mislabeled, it cannot be the mixed jar and must be, in reality, the nickel jar. Since the jar marked dimes cannot really be dimes, it must be the "mixed" jar. This leaves the third jar to be the dimes jar. You are probably thinking how simple the problem is— now that you have the solution. It does demonstrate a certain beauty of logical thinking.

THE FLIGHT OF THE BUMBLEBEE

Some problems present situations that lend themselves to very clever solutions. Oftentimes, after seeing the solution, we reflect over the solution with amazement. It is from such unusual approaches to a solution that one learns problem solving, since one of the most useful techniques in approaching a problem to be solved is to ask yourself: "Have I ever encountered such a problem before?" With this in mind, a problem with a very useful "lesson" is presented here. Don't be deterred by the relatively lengthy reading required to get through the problem. You will be delighted (and entertained) with the unexpected simplicity of the solution.

> **Two trains, serving the Chicago to New York route, a distance of 800 miles, start toward each other at the same time (along the same tracks). One train is traveling uniformly at 60 miles per hour and the other at 40 miles per hour. At the same time a bumblebee begins to fly from the front of one of the trains, at a speed of 80 miles per hour toward the oncoming train. After touching the front of this second train, the bumblebee reverses direction and flies toward the first train (still at the same speed of 80 miles per hour). The bumblebee continues this back-and-forth flying until the two trains collide, crushing the bumblebee. How many miles did the bumblebee fly before its demise?**

It is a natural inclination to want to find the individual distances that the bumblebee traveled. An immediate reaction is to set up an equation based on the famous (from high school mathematics) relationship: "rate times time equals distance." However, this back-and-forth path is rather difficult to determine; that is, it would require considerable calculation. Just the notion of having to do this can cause serious frustration. Do not allow this frustration to set in. Even if you were able to determine each of the parts of the bumblebee's flight, it would be still very difficult to solve the problem.

A much more elegant approach would be to look at the problem from a different point of view. We seek to find the *distance* the bumblebee traveled. If we knew the *time* the bumblebee traveled, we could determine the bumblebee's distance because we already know the *speed* of the bumblebee. Having two parts of the equation "rate × time = distance" will provide the third part. So having the *time* and the *speed* will yield the distance traveled—albeit in various directions.

The time the bumblebee traveled can be easily calculated, since it traveled the entire time the two trains were traveling toward each other (until they collided). To determine the time, t, that the trains traveled, we need to set up an equation as follows: The distance[10] traveled by the first train is $60t$ and the distance traveled by the second train is $40t$. The total distance the two trains traveled is 800 miles. Therefore, $60t + 40t = 800$, so $t = 8$ hours, which is also the time the bumblebee traveled. We can now find the distance the bumblebee traveled, by again using the relationship, rate × time = distance, which gives us $(8)(80) = 640$ miles.

---

10. The distance is equal to the product of the time and the speed.

It is important to stress how to avoid falling into the trap of always trying to do what the problem calls for directly. At times, a more circuitous method is much more efficient. Lots can be learned from this solution. You see, dramatic solutions are often more useful than traditional solutions, since they provide an opportunity to "think outside of the box."

UNDERSTANDING LIMITS

The concept of a limit is not to be taken lightly; it is very sophisticated and can easily be misinterpreted. Sometimes the issues surrounding the concept are quite subtle. Misunderstanding them can lead to some curious (or humorous, depending on your viewpoint) situations. This is evident in the following two illustrations. Don't be too upset by the conclusion that you will be led to reach—remember, this is for entertainment. Consider them separately and then notice their connection.

In figure 5-29, we have a set of uneven stairs, consisting of horizontal and vertical line segments (shown in bold).

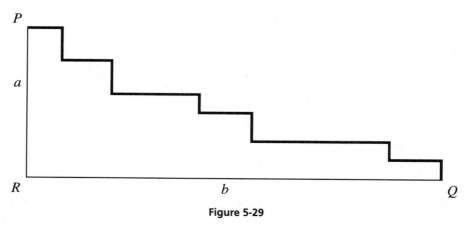

Figure 5-29

The sum of these bold segments ("stairs"), found by summing all the horizontal and all the vertical segments, is $a+b$, since it is the same as the vertical length, a, and the horizontal length, b. If the number of stairs increases, the sum is still $a+b$.

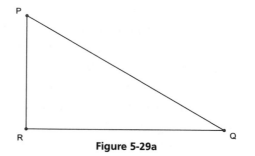

Figure 5-29a

The dilemma arises when we increase the stairs to a "limit" so that the set of stairs appears to be a straight line (figure 5-29a); in this case, it would be the hypotenuse of $\triangle PQR$. It would then appear that PQ has length $a+b$. Yet we know from the Pythagorean Theorem that $PQ = \sqrt{a^2 + b^2}$ and is *not* equal to $a+b$. So what's wrong?

Nothing is wrong! While the set consisting of the stairs does indeed approach closer and closer to the straight line segment PQ, it does *not* therefore follow that the *sum* of the bold (horizontal and vertical) lengths approaches the length of PQ, contrary to intuition. There is no contradiction here, only a failure on the part of our intuition.

Another way to "explain" this dilemma is to argue the following: As the "stairs" get smaller, they increase in number. In an extreme situation, we have 0-length dimensions (for the stairs) used an infinite number of times, which then leads to considering $0 \cdot \infty$, which is meaningless!

A similar situation arises with the following example. In figure 5-30, the smaller semi-circles extend from one end of the large semicircle's diameter to the other.

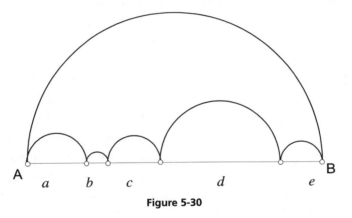

Figure 5-30

It is easy to show that the sum of the arc lengths of the smaller semicircles is equal to the arc length of the larger semicircle. We can write the sum of the smaller semicircles as:

$$\text{Sum} = \frac{\pi a}{2} + \frac{\pi b}{2} + \frac{\pi c}{2} + \frac{\pi d}{2} + \frac{\pi e}{2} = \frac{\pi}{2}(a+b+c+d+e) = \frac{\pi}{2}(AB)$$

which just happens to be the arc length of the larger semicircle. This may be surprisingly deceptive, but it is true. As a matter of fact, as we increase the number of smaller semi-circles (where, of course, they get smaller), the sum "appears" to be approaching the length of the segment AB, but, in fact, does not!

Again, the set consisting of the semicircles does indeed approach the length of the straight-line segment AB. It does *not* follow, however, that the *sum* of the semicircles approaches the *length* of the limit (in this case AB).

This "apparent limit sum" is absurd, since the shortest distance between points A and B is the length of segment AB, not the semicircle-arc AB (which, you will remember, equals the sum of the smaller semicircles). This important concept of limit is always difficult to conceive, yet it may be best explained with the help of such illustrations, so that future misinterpretations can be avoided.

WHERE IS THE MISSING MONEY?

Sometimes what appears to be true is not. We see this with some simple business dealings. Suppose there were two vendors on the street selling belts. Each had 120 belts to sell. One sold them at two for $5 and the other vendor sold the belts at three for $5. The first vendor sold all of his belts and took in $60 \cdot \$5 = \$300$, while the second vendor also sold all of his belts and took in $40 \cdot \$5 = \$200$. Together they brought in $500.

The next week they decided to combine their 240 belts and sell them at the rate of five for $10—which they felt was the combination of their two previous prices (i.e., two for $5 and three for $5). When they sold all their belts, they took in $48 \cdot \$10 = \$480$. What happened to the $20 difference?

It appears that everything was perfectly logical. When they calculated their average rate of selling belts, they found the rates of sale 2 for $5, or $\frac{2}{5}$, and 3 for $5, or $\frac{3}{5}$, over the same number of belts. What they should have done is to divide the total number of belts by the total number of dollars: $\frac{240}{500} = \frac{12}{25}$, which would then have given the correct rate at which they should have sold belts in the combined form. They sold the 240 belts in the second week at the rate of $\frac{240}{480} = \frac{1}{2}$. So, you can see the difference in the selling price. Such confused calculations can make us wonder about our arithmetic abilities, but regardless of whether or not we use a calculator, our reasoning must be correct.

WHEN "AVERAGES" ARE NOT AVERAGES

We must begin by familiarizing ourselves with a "baseball batting average." Most people, especially after trying to explain this concept, will begin to realize that it is not an average in the way they usually define an "average"—the arithmetic mean. It might be good to search the sports section of the local newspaper to find two baseball players who currently have the same batting average but who have achieved their respective batting average with a different number of hits. Let's use a hypothetical example here.

Consider two players: Simon and Miriam, each with a batting average of .667. Simon achieved his batting average by getting 20 hits for 30 at bats, while Miriam achieved her batting average by getting 2 hits for 3 at bats.

On the next day both perform equally, getting 1 hit for 2 at bats (for a .500 batting average). One might expect that they would then still have the same batting average at the end of the day. We can calculate their respective averages: Simon now has $20 + 1 = 21$ hits for $30 + 2 = 32$ at bats for a $\frac{21}{32} \approx .656$ batting average.

Miriam now has $2 + 1 = 3$ hits for $3 + 2 = 5$ at bats for a $\frac{3}{5} = .600$ batting average. Surprise! They do not have equal batting averages.

Suppose we consider the next day, where Miriam performs considerably better than Simon does. Miriam gets 2 hits for 3 at bats, while Simon gets 1 hit for 3 at bats. We shall now calculate their respective averages:

Simon has $21 + 1 = 22$ hits for $32 + 3 = 35$ at bats for a batting average of $\frac{22}{35} \approx .629$.

Miriam has $3 + 2 = 5$ hits for $5 + 3 = 8$ at bats for a batting average of $\frac{5}{8} = .625$.

Amazingly, despite Miriam's superior performance on this day, her batting average (which was the same as Simon's at the start) is still lower. The so-called batting average is really not an average in the true sense, but rather a "batting rate," and as you know, we cannot just add rates—they must be handled in a different way. In this situation, we need to consider averaging rates. These could be rates of purchase, or rates of speed, and so on.

Let's go right to another analogous problem: On Monday, a plane makes a round-trip flight from New York City to Washington and back with an average speed of 300 miles per hour. The next day, Tuesday, there is a wind of constant speed (50 miles per hour) and consistent direction (blowing from New York City to Washington). With the same speed setting as on Monday, this same plane makes the same round-trip on Tuesday. Will the Tuesday trip require more time, less time, or the same amount of time as the Monday trip?

Bear in mind that the only thing that has changed is the support and hindrance of the wind. All other controllable factors are the same: distances, speed regulation, the airplane's condition, and so on. An expected response is that the two round-trip flights ought to be the same, especially since the same wind is helping and hindering two equal legs of a round-trip flight.

However, we shall digress to an entirely different situation, letting this problem lie unresolved for a while. Consider the situation about the grade a student deserves who scored 100% on nine of ten tests in a semester and on one test scored only 50%. Would it be fair to assume that this student's performance for the term was 75% (i.e., $\frac{100+50}{2}$)? The reaction to this suggestion will tend toward applying appropriate weight to the two scores in consideration. The 100% was achieved nine times as often as the 50%, and therefore ought to get the appropriate weight. Thus, a proper calculation of the student's average ought to be $\frac{9 \cdot 100 + 50}{10} = 95$. This clearly appears more just!

Now, how might this relate to the airplane trip? The realization that the two legs of the "wind-trip" require different amounts of time should lead to the notion that the two speeds of this trip cannot be weighted equally, as they were done for different lengths of time. Therefore, the time for each leg should be calculated and then appropriately apportioned to the related speeds. Again, we shall use the formula: rate times time equals distance, or in this case, time is equal to the distance divided by the rate.

We first find the times $(t_1$ and $t_2)$ of the two legs of the round-trip in wind: $t_1 = \dfrac{d}{350}$ and $t_2 = \dfrac{d}{250}$.

The total time for the wind round-trip is: $t = \dfrac{d}{350} + \dfrac{d}{250}$.

The "total rate" or rate for the wind round-trip (which is really the average rate) is:
$r = \dfrac{2d}{\dfrac{d}{250} + \dfrac{d}{350}} = \dfrac{(2)(350)(250)}{250 + 350} \approx 291.67$, which is slower than the no-wind trip, and hence took more time.

This average of rates—in this case ≈ 291.67—is called the *harmonic mean*[11] between the two speeds of 350 mph and 250 mph.

MATHEMATICAL PARADOXES

Imagine that in algebra we could show something to be true that in fact was not true. If this were really the case, then we would lose all confidence in this most popular of mathematical languages. You might find that we would conveniently seek to avoid such situations that were false. In that spirit, follow along with this "proof" that $2 = 1$, and see if you can identify the error.

Let	$a = b$
Multiply both sides by a:	$a^2 = ab$
Subtract b^2 from both sides:	$a^2 - b^2 = ab - b^2$
Factor:	$(a+b)(a-b) = b(a-b)$
Divide both sides by $(a-b)$:	$(a+b) = b$
Since $a = b$, then	$2b = b$
Divide both sides by b:	$2 = 1$

11. The harmonic mean is the reciprocal of the arithmetic mean of the reciprocals of the quantities being considered. The harmonic mean for a and b is $\dfrac{1}{\dfrac{\dfrac{1}{a} + \dfrac{1}{b}}{2}} = \dfrac{2}{\dfrac{1}{a} + \dfrac{1}{b}} = \dfrac{2ab}{a+b}$, and for three numbers, a, b, and c, the

harmonic mean is $\dfrac{1}{\dfrac{\dfrac{1}{a} + \dfrac{1}{b} + \dfrac{1}{c}}{3}} = \dfrac{3}{\dfrac{1}{a} + \dfrac{1}{b} + \dfrac{1}{c}} = \dfrac{3abc}{ab + ac + bc}$.

The astute reader will notice that in step 5 we divided by $a - b$, which is zero, since $a = b$. This violates the definition of division, which prohibits division by 0. You may ask why division by 0 is not defined. The obvious answer is that if it were permissible, then the dilemma found in the "proof" (above) would hold true, and $2 = 1$. From your knowledge of arithmetic, you should realize that if $\frac{x}{0} = y$ (where $x \neq 0$), then $y \cdot 0 = x$, but there is no value y for which this can be true. Thus, division by zero leads to "inconsistencies" and must remain undefined. There are a number of other demonstrations of weird results where this definition is not heeded.

1. To "prove" that any two unequal numbers are equal. Assume that $x = y + z$, and x, y, z are positive numbers. This implies $x > y$. Multiply both sides by $x - y$. Then $x^2 - xy = xy + xz - y^2 - yz$. Subtract xz from both sides: $x^2 - xy - xz = xy - y^2 - yz$. Factoring, we get $x(x - y - z) = y(x - y - z)$. Dividing both sides by $(x - y - z)$ yields $x = y$. Thus x, which was assumed to be greater than y, has been shown to equal y. The fallacy occurs in the division by $(x - y - z)$, which is equal to zero.

2. To "prove" that all positive whole numbers are equal. By doing algebraic division, we have, for any value of x:

$$\frac{x - 1}{x - 1} = 1$$

$$\frac{x^2 - 1}{x - 1} = x + 1$$

$$\frac{x^3 - 1}{x - 1} = x^2 + x + 1$$

$$\frac{x^4 - 1}{x - 1} = x^3 + x^2 + x + 1$$

$$\vdots$$

$$\frac{x^n - 1}{x - 1} = x^{n-1} + x^{n-2} + \cdots + x^2 + x + 1$$

Letting $x = 1$ in all of these identities, the right side then assumes the values $1, 2, 3, 4, \ldots, n$. The left side of each of the identities[12] assumes the value $\frac{0}{0}$ when $x = 1$. This problem serves as evidence that $\frac{0}{0}$ is meaningless.

---

12. An *identity* is an equation that is true for all values of the variable.

Another often-overlooked definition is dramatized by the following demonstration: You will remember that $\sqrt{a}\sqrt{b}=\sqrt{ab}$; for example, $\sqrt{2}\cdot\sqrt{5}=\sqrt{2\cdot5}=\sqrt{10}$. But this can also result in a dilemma when we consider $\sqrt{-1}\sqrt{-1}=\sqrt{(-1)(-1)}=\sqrt{1}=1$, since we also know that $\sqrt{-1}\sqrt{-1}=\left(\sqrt{-1}\right)^{2}=-1$. It therefore may be concluded that $1=-1$, since both equal $\sqrt{-1}\sqrt{-1}$. Can you explain the error? In short, we cannot apply the ordinary rules for multiplication of radicals to imaginary numbers.[13] $\sqrt{a}\sqrt{b}=\sqrt{ab}$ is valid only when the values of a and b are not negative.

This dilemma caused by incorrectly treating the product of two imaginary numbers can further be seen with the following demonstration:

Consider the "proof" that can be used to show $-1=+1$:

$$\sqrt{-1}=\sqrt{-1}$$
$$\sqrt{\frac{1}{-1}}=\sqrt{\frac{-1}{1}}$$
$$\frac{\sqrt{1}}{\sqrt{-1}}=\frac{\sqrt{-1}}{\sqrt{1}}$$
$$\sqrt{1}\sqrt{1}=\sqrt{-1}\sqrt{-1}$$
$$1=-1$$

If you replace i for $\sqrt{-1}$ and -1 for i^2, it's easy to see where the flaw occurs. Remember that $i^{2}=-1$.

With each definition violation, awkward or weird results will occur. This, in itself, justifies the definitions.

What is the sum of the infinitely long series: $1-1+1-1+1-1+1-1+\cdots$? This is a rather tricky question and will lead us to a dilemma. First, we can use the following relationship: $\frac{1}{x+1}=1-x+x^{2}-x^{3}+x^{4}-x^{5}+\cdots$.

If we let $x=1$, then the above equation reads: $\frac{1}{2}=1-1+1-1+1-1+1-1+\cdots$, and so our sought-after answer seems to be $\frac{1}{2}$.

However, if we consider the following true statement:

$\frac{1}{x^{2}+x+1}=1-x+x^{3}-x^{4}+x^{6}-x^{7}+\cdots$, then, when we let $x=1$,

we get $\frac{1}{3}=1-1+1-1+1-1+1-1+\cdots$

Does this mean that $\frac{1}{2}=\frac{1}{3}$?

---

13. The square root of a negative number is called an *imaginary number*.

We can even take this further. Consider the equation:

$$\frac{1}{x^3+x^2+x+1}=1-x+x^4-x^5+x^8-x^9+\cdots$$

This time, when we let $x=1$, we get $\frac{1}{4}=1-1+1-1+1-1+1-1+\cdots$, which would imply that $\frac{1}{2}=\frac{1}{3}=\frac{1}{4}$.

If we continue this pattern, we would have to conclude that $\frac{1}{2}=\frac{1}{3}=\frac{1}{4}=\cdots=\frac{1}{n}$, which is clearly absurd!

The "confusion" lies in the fact that the value of $1-1+1-1+1-1+1-1+\cdots$ oscillates between 0 and 1.

We can see this by how we group the terms of this infinite series. If we group it as:

$$(1-1)+(1-1)+(1-1)+(1-1)+\cdots$$
$$=0+0+0+0+0+\cdots$$
$$=0$$

On the other hand, if we group the terms of this infinite series as:

$$1-1+1-1+1-1+1-1+\cdots$$
$$=1-(1-1)-(1-1)-(1-1)-\cdots$$
$$=1-0-0-0-0-0-\cdots$$
$$=1$$

In short, the "error" above occurred because the series $1-1+1-1+1-1+1-1+\cdots$ does not have a definite sum.

Sometimes physical observations are very difficult to explain, and can even be paradoxical. For example, we know that when a circle rolls on a line and makes one complete revolution, then it has traveled the distance equal to the length of its circumference. In figure 5-31, when the larger circle travels from point D to point E, it will have traveled the distance DE, which is equal to the circumference of the larger circle. When you consider the two concentric circles rolling, whose circumferences are not equal, we wonder how the smaller circle will have traveled one large-circle-circumference length at the same time as the larger circle traveled a longer distance. This may be seen in figure 5-31. AB is equal to DE, yet it cannot be the length of the circumference of the smaller circle. How is this possible?

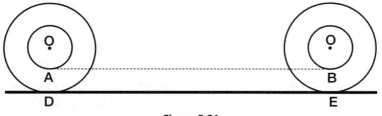

Figure 5-31

CAN A SEQUENCE OF NUMBERS DEFY PHYSICS?

The sequence of even numbers 2, 4, 6, 8, 10, 12, $\cdots$ is a very popular one in mathematics. If we take the reciprocals of each of these numbers, we get a sequence that has a special property—one that reminds us of a harmonic sequence. It looks like this: $\frac{1}{2}, \frac{1}{4}, \frac{1}{6}, \frac{1}{8}, \frac{1}{10}, \frac{1}{12}, \cdots$.

You will notice that these unit fractions have successive even number denominators.

The series of unit fractions $\frac{1}{2}, \frac{1}{4}, \frac{1}{6}, \frac{1}{8}, \frac{1}{10}, \frac{1}{12}, \cdots$ is part of the harmonic sequence $1, \frac{1}{2}, \frac{1}{3}, \frac{1}{4}, \frac{1}{5}, \cdots$.

There must be something harmonic about this sequence of numbers. If you take guitar strings with the same tension and set them up with lengths represented by these fractions and then strum them together, you will get a harmonious sound! Now let's see what this sequence of numbers might show us physically.

In figure 5-32, you will see some domino tiles that are stacked in a way that you would expect them to tip over.

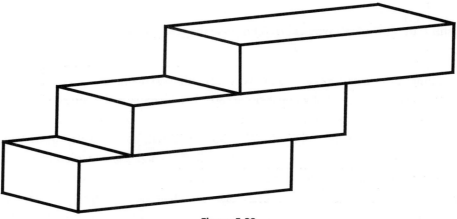

Figure 5-32

It appears that the center of gravity of the top tile is beyond the bottom tile. Is this sort of stacking possible?

Amazingly enough, this is possible, but it would require an unusual stacking of the tiles. To get the right tile spacing for the stack, we would approach this somewhat unusually. The method of stacking these tiles—or at least determining the proper stacking method—would be to work backward, namely, starting with the top tile and then working downward.

Our scheme for the domino stacking would be to extend each tile a certain fraction of its entire length beyond the tile on which it rests. The extension fractions would be from the (harmonic) sequence of unit fractions: $\frac{1}{2}, \frac{1}{4}, \frac{1}{6}, \frac{1}{8}, \frac{1}{10}, \frac{1}{12}, \dots$.

In figure 5-33, you will notice how each tile is placed so that the specific unit fraction of its length is extended over the tile directly beneath it.

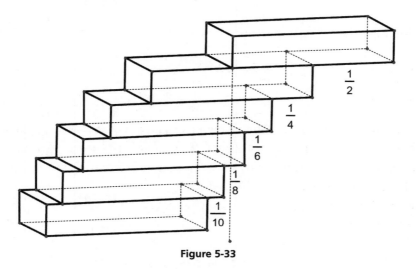

Figure 5-33

A side view of the tile stack in figure 5-33 is shown in figure 5-34.

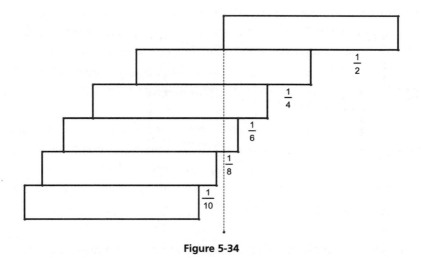

Figure 5-34

Since the sum of the overhang pieces gets larger, the center of gravity will constantly move farther away from the bottom tile. In mathematics, we say the sum, $\frac{1}{2} + \frac{1}{4} + \frac{1}{6} + \frac{1}{8} + \cdots$ diverges, or keeps on getting larger.

Number domino from top	Extension length beyond previous domino	Sum of the overhang		
1	$\dfrac{1}{2}$	$\dfrac{1}{2}$	$=\dfrac{1}{2}$	$= 0.5$
2	$\dfrac{1}{4}$	$\dfrac{1}{2}+\dfrac{1}{4}$	$=\dfrac{3}{4}$	$= 0.75$
3	$\dfrac{1}{6}$	$\dfrac{1}{2}+\dfrac{1}{4}+\dfrac{1}{6}$	$=\dfrac{11}{12}$	≈ 0.9166
4	$\dfrac{1}{8}$	$\dfrac{1}{2}+\dfrac{1}{4}+\dfrac{1}{6}+\dfrac{1}{8}$	$=\dfrac{25}{24}$	≈ 1.0416
5	$\dfrac{1}{10}$	$\dfrac{1}{2}+\dfrac{1}{4}+\dfrac{1}{6}+\dfrac{1}{8}+\dfrac{1}{10}$	$=\dfrac{137}{120}$	≈ 1.1416
6	$\dfrac{1}{12}$	$\dfrac{1}{2}+\dfrac{1}{4}+\dfrac{1}{6}+\dfrac{1}{8}+\dfrac{1}{10}+\dfrac{1}{12}$	$=\dfrac{49}{40}$	$= 1.225$
...				
100	$\dfrac{1}{200}$	$\displaystyle\sum_{n=1}^{100}\dfrac{1}{2n}$		≈ 2.5936
...				
1,000	$\dfrac{1}{2,000}$	$\displaystyle\sum_{n=1}^{1000}\dfrac{1}{2n}$		≈ 3.7427

Figure 5-35

As we mentioned above, the logical way to build this stack of dominos is to begin at the bottom, but here we must (unusually) start at the top and work our way down. After we reach the fifth domino from the bottom, we find that this entire tile is well beyond the bottom tile (figure 5-34). It is unbelievable that this tile can then be supported by the bottom tile.

Were we to extend this to 100 dominos, we would find the top domino to be extended more than $2\frac{1}{2}$ dominolengths (i.e., from the chart in figure 5-35: ≈ 2.5936) over the bottom one.

How can this be explained? A domino's middle is its center of gravity. It is expected that as long as the top domino's center of gravity is above a portion of the domino beneath it, it will keep from tipping over. The limiting condition is when that center of gravity is at the edge of the domino beneath it. (This is shown with the black dot in figure 5-36.)

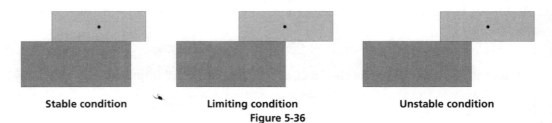

Stable condition   Limiting condition   Unstable condition

Figure 5-36

With each additional domino tile, the center of gravity of the entire stack (depicted by the white dot in figure 5-37) moves farther out. The fourth tile leads us to our unexpected conclusion, since the next tile (the fifth tile) then extends beyond the base tile.

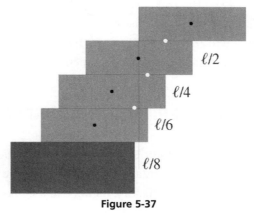

$\ell/2$

$\ell/4$

$\ell/6$

$\ell/8$

Figure 5-37

This "counterintuitive condition," based on a very common harmonic sequence, is unbelievable, but true!

A NICE LITTLE SURPRISE

Most people remember the Pythagorean Theorem from their school days. And, even if they do not remember the theorem, they will recall the equation $a^2 + b^2 = c^2$. Remember, the lengths of the sides of a right triangle, *a*, *b*, and *c*, have this relationship. The most common application of this equation is the relationship: $3^2 + 4^2 = 5^2$. This is the only set of three *consecutive* numbers that has this relationship and forms what we call a Pythagorean triple.[14] Surprisingly, we can extend this set of three consecutive numbers—each taken to the second power—and extend it to the sum of three consecutive numbers—each taken to the third power. Yes, it just happens also to be true that $3^3 + 4^3 + 5^3 = 6^3$. And so we have another little nugget in our collection of mathematical surprises.

Besides the obvious, namely, that the Pythagorean Theorem relates to the side lengths of right triangles, many clever proofs of this relationship exist. Actually, well over **520** proofs of this theorem have already been published. Every high school geometry textbook has at least one of these proofs.

There are lots of interesting relationships that can be found among the Pythagorean triples. For example, the product of any primitive Pythagorean triple is always divisible

---

14. A *primitive Pythagorean triple* is a set of three integers with no common factor, where the sum of the squares of two of them is equal to the square of the third number.

by 60. The investigation of types and the nature of these Pythagorean triples are open to much further investigation. For instance, some of these triples contain a pair of consecutive numbers, such as: (3, 4, 5), (5, 12, 13), (7, 24, 25), (9, 40, 41), (11, 60, 61), (13, 84, 85), (15, 112, 113), (17, 144, 145), (19, 180, 181), (21, 220, 221), and even larger ones such as: (95, 4512, 4513). We leave you with this brief introduction to the Pythagorean Theorem with the hope that you will investigate these Pythagorean triples further. The journey through their many relationships and properties is practically boundless!

THE VALUE OF π: FROM PRIMITIVE TO PRESENT

You'll be surprised to know that for centuries scholars believed that 3 was the value that π had in biblical times. For many years, virtually all the books on the history of mathematics stated that in its earliest manifestation in history, namely, in the Old Testament of the Bible, the value of π was given as 3. Yet recent "detective work" shows otherwise.[15]

One always relishes the notion that a hidden code can reveal long-lost secrets. Such is the case with the common interpretation of the value of π in the Bible. There are two places in the Bible where the same sentence appears, identical in every way except for one word, which is spelled differently in the two citations. The description of a pool, or fountain, in King Solomon's temple is referred to in the passages that may be found in 1 Kings 7:23 and 2 Chronicles 4:2, and reads as follows:

> And he made the molten sea[16] of ten cubits from brim to brim, round in compass, and the height thereof was five cubits; and a **line measure** of thirty cubits did compass it round about.

The circular structure described here is said to have a circumference of 30 cubits[17] and a diameter of 10 cubits. From this we notice that the Bible has $\pi = \frac{30}{10} = 3$. This is obviously a very primitive approximation of π. A late eighteenth-century mathematician and rabbi, Elijah of Vilna (1720–1797),[18] was one of the great modern biblical scholars, who earned the title "Gaon of Vilna" (meaning brilliance of Vilna). He came up with a remarkable discovery, one that would make most history of mathematics books faulty if they say that the Bible approximated the value of π as 3. Elijah of Vilna noticed that the Hebrew word for "line measure" was written differently in each of the two biblical passages mentioned above.

---

15. A. S. Posamentier and I. Lehmann, π: *A Biography of the World's Most Mysterious Number* (Amherst, NY: Prometheus Books, 2004).
16. The "molten sea" was a gigantic bronze vessel for ritual ablutions in the court of the First Temple (966–955 BCE). It was supported on the backs of twelve bronze oxen (volume ≈ 45,000 liters).
17. A *cubit* is the length of a person's fingertip to his elbow.
18. In those days Vilna was in Poland, while today the town is named Vilnius and is in Lithuania.

In 1 Kings 7:23 it was written as קוה, whereas in 2 Chronicles 4:2 it was written as קו—but read from right to left.

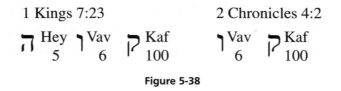

1 Kings 7:23   2 Chronicles 4:2

ה Hey   ו Vav   ק Kaf   ו Vav   ק Kaf
   5   6   100   6   100

Figure 5-38

Elijah applied the ancient biblical analysis technique (still used by Talmudic scholars today) called *gematria*, where the Hebrew letters are given their appropriate numerical values according to their sequence in the Hebrew alphabet, to the two spellings of the word for "line measure." He found the following. The letter values are: ק = 100, ו = 6, and ה = 5. Therefore, the spelling for "line measure" in 1 Kings 7:23 is קוה = 5 + 6 + 100 = 111, while in 2 Chronicles 4:2 the spelling קו = 6 + 100 = 106. Using gematria in an accepted way, he then took the ratio of these two values: $\frac{111}{106} = 1.0472$ (to four decimal places), which he considered the necessary "correction factor." By multiplying the Bible's apparent value (3) of π by this "correction factor," one would get 3.1416, which is π correct to four decimal places! "Wow!" is a common reaction. Such accuracy is quite astonishing for ancient times.

Just to put things into proper perspective, today—with the aid of a supercomputer—we have the value of π calculated to 1.24 trillion decimal place accuracy! This value was found in December 2002, by Professor Yasumasa Kanada (a longtime pursuer of π) and nine others at the Information Technology Center at Tokyo University. They accomplished this feat with a Hitachi SR8000 supercomputer, which is capable of doing 2 trillion calculations per second.

The number π harbors endless surprises beyond determining its value.[19] This is a good springboard to pursue further investigation.

Finding the value of π is only one of the many delights that mathematics has to offer. It is our hope that in sampling the many exciting offerings in mathematics, your curiosity will be piqued and lead you to further explorations.

ARCHIMEDES' SECRET

On October 29, 1998, Christie's auction house in New York sold a small book from the thirteenth century for $2.2 million. On the surface it appeared to be a prayer book, but that is not the reason for its extraordinarily high value. This book is a collection of parchment

---

19. We recommend reading the book by A. S. Posamentier and I. Lehmann, π: *A Biography of the World's Most Mysterious Number* (Amherst, NY: Prometheus Books, 2004).

sheets that had been chemically erased by monks during the thirteenth century and reused to produce a prayer book. Although it had some damage, through modern technology we were able to detect that its pages were originally written by Archimedes (ca. 287–212 BCE) and it is referred to today as the Codex C. It was truly a sensation when, in 1998, this long-lost book reappeared. It is what is believed to be one of three such books about mathematics written by Archimedes. The key to this wonderful discovery was the Stanford University Linear Accelerator Center, which was able to enable us to see through the over-written material.

Among the many treasures in this Codex C is a discussion of the Stomachion.[20] This is perhaps one of our earliest-known puzzle games, which consists of fourteen pieces that are either triangles, quadrilaterals, or pentagons (see figure 5-39).

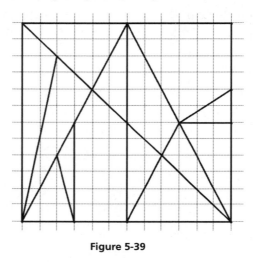

Figure 5-39

The object of the puzzle is to place the fourteen pieces in such a way that the resulting figure is a square. Although this may appear analogous to the modern tangram game, it is far more challenging. It has been known since 1906[21] that Archimedes had grappled with a Stomachion. However, it was not until the discovery of this book that we realized that he actually concerned himself with a field of combinatorics[22] by considering the number of possible arrangements of these fourteen puzzle pieces to make a square. Bear in mind that the field of combinatorics was officially not introduced into our current mathematical knowledge until the seventeenth century by Pierre de Fermat (1607–1665) and Blaise

---

20. This is from the Greek word for stomachache, something that could be evoked from frustration in seeking solutions to the puzzle it names.
21. Reviel Netz and William Noel, *The Archimedes Codex: Revealing the Secrets of the World's Greatest Palimpsest* (London: Weidenfeld & Nicolson, 2007).
22. This is a part of the field of probability and considers the number of ways a given set of objects can be arranged.

Pascal (1623–1662). It turns out that Archimedes' question has been answered in that the Stomachion puzzle has 17,152 possible solutions, and even when we take symmetry into account, it still has 536 solutions. In figures 5-40 and 5-41 you will see two solutions to the problem and notice that when we put the fourteen parts together, none of them overlap each other.

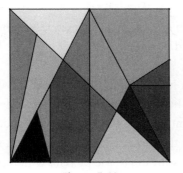

Figure 5-40

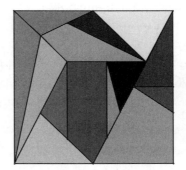

Figure 5-41

THE ETERNITY PUZZLE

As problems get solved, we continue to look to increase the challenge. The *Eternity puzzle* is an outgrowth of the Stomachion, requiring the creation of a dodecagon (a twelve-sided polygon) from a set of 209 tiles. The challenges initiated by the Stomachion continue boundlessly still today.

The Eternity puzzle—a "grandchild" of the Stomachion of Archimedes—proves that one can win lots of money by successfully completing a jigsaw puzzle. On the other hand, the "great-grandchild" of the Stomachion of Archimedes, often referred to as *Eternity II*, has to date not yet been solved. Two young mathematicians from Cambridge, Alex Selby and Oliver Riordan, won one million British pounds sterling for the solution of the Eternity puzzle.

Instead of the 14 parts that have to be put together in the Stomachion to form a square, the Eternity puzzle consists of 209 parts that must be placed to fit into an enclosed dodecagon—one that is almost regular.[23]

In June 1999, the British company Racing Champions Ltd. marketed this puzzle, which was developed by Christopher Monckton (1952–)[24] from Aberdeenshire, Scotland. The components of this puzzle are irregular polygons, which have between six and eleven sides and are comprised of "drafter's triangles."[25]

---

23. A *regular* polygon is one that has all sides of the same length and all angles of equal measure.
24. Christopher Walter Monckton, 3rd. Viscount Monckton of Brenchley.
25. "Drafter's triangles" are triangles with the angles 30°, 60°, and 90°.

For example, figure 5-42 shows three pieces that fit together properly. The puzzle begins with the dilemma that each piece seems to be able to fit everywhere. Neither the shape nor the color gives enough information about where a piece should be placed in order to complete the puzzle. Since all the pieces have the same color, it is a great help to color them as indicated in figure 5-42. Furthermore, it is helpful to triangulate the pieces with the right triangles shown in figure 5-42.

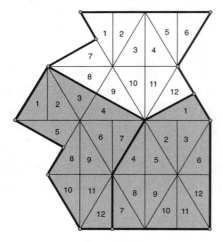

Figure 5-42

Rumors abounded that this puzzle was made as revenge toward the creator's former boss, the former British prime minister Margaret Thatcher. When Monckton left his position as one of her advisors in 1986, he gave her a twelve-piece puzzle made out of silver. Apparently, after a few years of unsuccessful attempts to do the puzzle, Thatcher sent him a letter requesting the solution to the puzzle, which seemed to have brought Monckton to the idea of marketing the puzzle and offering a one million pound sterling award for it solution. He expected to get a solution within three years. Yet there were those who did not expect to see a solution in their lifetime.

A computer solution seemed to be the expected path to success—which after barely one year (on May 15, 2000) seemed to have materialized through the efforts of Selby and Riordan. They collected their prize money in September of that year.

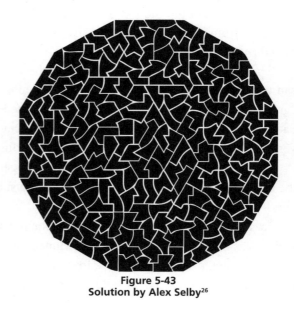

Figure 5-43
Solution by Alex Selby[26]

26. Used with the permission of Alex Selby. The pieces are copyright Christopher Monckton 1999 and are reproduced here with his permission.

The second person to solve the puzzle was the German mathematician Günter Stertenbrink in July 2000. Interestingly, his solution appeared entirely different than that of the two Britons. The number of possible solutions to the puzzle is still unknown, but the current thinking is that there are about 10^{100} (or one googol) such solutions.

The successor puzzle, now called Eternity II, was developed by Christopher Monckton on June 28, 2007—this time in cooperation with the winners of the first prize money, Selby and Riordan. The prize money was two million US dollars, and the puzzle is a worthy successor of the Stomachion because it has 256 squares with sides of varying colors and designs. More about this puzzle can be found at www.eternityii.com.

As opposed to the Eternity puzzle, there probably are only a few solutions to Eternity II. Monckton is quoted as follows in the *Times*: "Our calculations are that if you used the world's most powerful computer and let it run from now until the projected end of the universe, it might not stumble across one of the solutions."[27]

Although many of our puzzle games today emanate from earlier recreations, no one anticipated that as far back as Archimedes would a mathematician have concerned himself with such a sophisticated analysis of a puzzle. Amazements in mathematics seemed to be limitless!

---

27. *Times* online, December 4, 2005.

AFTERWORD

By Herbert A. Hauptman

Mathematics was always my favorite subject in school. Having been told in Stockholm at the award ceremony that I was the first mathematician to win a Nobel Prize gave me a chance to elevate mathematics to the level it should have, and finally have it recognized by the Nobel Prize. Remember, for inexplicable reasons, Alfred Nobel chose not to include mathematics among the categories for award. Yet one of the greatest mathematicians of all time, Carl Friedrich Gauss, referred to mathematics as the "queen of science."

Exactly where my love for mathematics began is not clear to me. Surely, I always excelled in the subject from my earliest school days. I do, however, remember some episodes that reinforced my good feelings about mathematics. One such occurred when I was a student at (the original) Townsend Harris High School in New York City—a school for gifted students, whose real prize was guaranteed admission (after completing three years) to the City College of New York, and a free and wonderful university education. This was particularly prized as it was in the midst of the Great Depression.

During this time I remember one mathematical "encounter" that highlighted an amazing fact about numbers, which I was able to use to my advantage and clearly impress my friends. One of my classmates posed a problem that had just about everyone stumped for days. When the problem finally reached me, I somehow got an inspiration that allowed me to solve the problem almost immediately. The problem is simple to state (that's what usually makes a problem attractive), and therefore appears "harmless." Yet the solution evaded my classmates, who were quite gifted in most subjects. I used a neat and surprising fact about numbers to my advantage.

It is well known that any integer can be expressed as the sum of powers of 2. We see this when we write numbers in binary notation—the basic tool of computers. This would not be of much help here, since we would need increasingly more powers of 2 as the numbers we wish to represent get larger. However, a less well-known fact is that any integer can be expressed as the sum or difference of powers of 3. It is this fact that was the key to the solution of the problem that vexed my high school friends.

So now let me present this problem and show you how this latter fact will lead us to a neat solution. The problem is:

> What is the least number of integer-pound weights needed to weigh an item weighing up to n-pounds on a balance scale, if you are allowed to place the weights on either side of the scale—that is, with, or separate from, the item to be weighed?

For example, to weigh an 11-pound weight, we could place a 9-pound weight and a 3-pound weight on one side of the balance scale and a 1-pound weight along with the item to be weighed on the other side of the scale. If the balance is even, then the item is truly 11 pounds. Naturally, it could be done with fewer weights, such as merely placing an 11-pound weight on one side to balance with the item to be weighed at 11 pounds. However, if we use this reasoning, then we would need n different weights to weigh items up to n pounds—clearly not the fewest number of weights!

To narrow down the number of weights required, I used the fact mentioned earlier that any integer can be expressed as the sum or difference of powers of 3. These are: $3^0, 3^1, 3^2, 3^3, 3^4, \cdots$, or 1, 3, 9, 27, 81 $\cdots$.

Let's see how this might be used for the first several weighings.

In the chart you will notice that to weigh items up to 4 pounds, we require only one or two weights (of 1 and 3 pounds). For items up to 13 pounds, we did not require more than three weights (of 1, 3, and 9 weights). Weighings up to 40 pounds require at most four weights. These are 1, 3, 9, and 27. You can see that weighings up to $1 + 3 + 9 + 27 + 81 = 121$ pounds will not require more than five weights: 1, 3, 9, 27, and 81, pounds. The weights are all powers of 3.

So we can then answer the original question of the problem. For weighing an item of $(1 + 3 + 9 + 27 + \cdots + 3^n)$ pounds, we will need at most $(n + 1)$ integer weights. This strategy also appears to give the minimum number of standard weights of unit measure (in powers of 3) needed to weigh objects of unknown weights. However, there can be internal ranges of weights for which the minimum would be lower.

Applying this nifty fact about numbers—that any integer can be expressed as the sum or difference of powers of 3—to this seemingly unrelated problem so impressed my high school classmates that I was even further motivated to pursue other mathematical investigations—some of which can be found in this nugget-filled book of mathematical amazements and surprises.

Weight of item (in bold)	Sum or difference of powers of 3 equal to the weight	Left side of balance scale	Right side of balance scale	Number of weights needed
1	1	1	**1**	1
2	3 − 1	3	**2**, 1	2
3	3	3	**3**	1
4	3 + 1	3, 1	**4**	2
5	9 − 3 − 1	9	**5**, 3, 1	3
6	9 − 3	9	**6**, 3	2
7	9 − 3 + 1	9, 1	**7**, 3	3
8	9 − 1	9	**8**, 1	2
9	9	9	**9**	1
10	9 + 1	9, 1	**10**	2
11	9 + 3 − 1	9, 3	**11**, 1	3
12	9 + 3	9, 3	**12**	2
13	9 + 3 + 1	9, 3, 1	**13**	3
14	27 − 9 − 3 − 1	27	**14**, 9, 3, 1	4
15	27 − 9 − 3	27	**15**, 9, 3	3
16	27 − 9 − 3 + 1	27, 1	**16**, 9, 3	4
17	27 − 9 − 1	27	**17**, 9, 1	3
18	27 − 9	27	**18**, 9	2
19	27 − 9 + 1	27, 1	**19**, 9	3
20	27 − 9 + 3 − 1	27, 3	**20**, 9, 1	4
21	27 − 9 + 3	27, 3	**21**, 9	3
22	27 − 9 + 3 + 1	27, 3, 1	**22**, 9	4
23	27 − 3 − 1	27	**23**, 3, 1	3
24	27 − 3	27	**24**, 3	2
25	27 − 3 + 1	27, 1	**25**, 3	3
26	27 − 1	27	**26**, 1	2
27	27	27	**27**	1
28	27 + 1	27, 1	**28**	2
29	27 + 3 − 1	27, 3	**29**, 1	3
30	27 + 3	27, 3	**30**	2
31	27 + 3 + 1	27, 3, 1	**31**	3
32	27 + 9 − 3 − 1	27, 9	**32**, 3, 1	4
33	27 + 9 − 3	27, 9	**33**, 3	3
34	27 + 9 − 3 + 1	27, 9, 1	**34**, 3	4
35	27 + 9 − 1	27, 9	**35**, 1	3
36	27 + 9	27, 9	**36**	2
37	27 + 9 + 1	27, 9, 1	**37**	3
38	27 + 9 + 3 − 1	27, 9, 3	**38**, 1	4
39	27 + 9 + 3	27, 9, 3	**39**	3
40	27 + 9 + 3 + 1	27, 9, 3, 1	**40**	4

ACKNOWLEDGMENTS

The authors acknowledge the outstanding editorial assistance provided by Peggy Deemer and Linda Greenspan Regan as well as the careful proofreading by Peter Poole. We also thank Dr. Elke Warmuth of Humboldt University in Berlin for her assistance with stochastics.

Appendix

LIST OF TRIANGULAR, SQUARE, AND CUBIC NUMBERS

n	Triangular numbers (t_n)	Square numbers (n^2)	Cubic numbers (n^3)
1	1	1	1
2	3	4	8
3	6	9	27
4	10	16	64
5	15	25	125
6	21	36	216
7	28	49	343
8	36	64	512
9	45	81	729
10	55	100	1,000
11	66	121	1,331
12	78	144	1,728
13	91	169	2,197
14	105	196	2,744
15	120	225	3,375
16	136	256	4,096
17	153	289	4,913
18	171	324	5,832
19	190	361	6,859
20	210	400	8,000

n	Triangular numbers (t_n)	Square numbers (n^2)	Cubic numbers (n^3)
21	231	441	9,261
22	253	484	10,648
23	276	529	12,167
24	300	576	13,824
25	325	625	15,625
26	351	676	17,576
27	378	729	19,683
28	406	784	21,952
29	435	841	24,389
30	465	900	27,000
31	496	961	29,791
32	528	1,024	32,768
33	561	1,089	35,937
34	595	1,156	39,304
35	630	1,225	42,875
36	666	1,296	46,656
37	703	1,369	50,653
38	741	1,444	54,872
39	780	1,521	59,319
40	820	1,600	64,000
41	861	1,681	68,921
42	903	1,764	74,088
43	946	1,849	79,507
44	990	1,936	85,184
45	1,035	2,025	91,125
46	1,081	2,116	97,336
47	1,128	2,209	103,823
48	1,176	2,304	110,592
49	1,225	2,401	117,649
50	1,275	2,500	125,000
51	1,326	2,601	132,651
52	1,378	2,704	140,608
53	1,431	2,809	148,877
54	1,485	2,916	157,464
55	1,540	3,025	166,375
56	1,596	3,136	175,616
57	1,653	3,249	185,193
58	1,711	3,364	195,112

n	Triangular numbers (t_n)	Square numbers (n^2)	Cubic numbers (n^3)
59	1,770	3,481	205,379
60	1,830	3,600	216,000
61	1,891	3,721	226,981
62	1,953	3,844	238,328
63	2,016	3,969	250,047
64	2,080	4,096	262,144
65	2,145	4,225	274,625
66	2,211	4,356	287,496
67	2,278	4,489	300,763
68	2,346	4,624	314,432
69	2,415	4,761	328,509
70	2,485	4,900	343,000
71	2,556	5,041	357,911
72	2,628	5,184	373,248
73	2,701	5,329	389,017
74	2,775	5,476	405,224
75	2,850	5,625	421,875
76	2,926	5,776	438,976
77	3,003	5,929	456,533
78	3,081	6,084	474,552
79	3,160	6,241	493,039
80	3,240	6,400	512,000
81	3,321	6,561	531,441
82	3,403	6,724	551,368
83	3,486	6,889	571,787
84	3,570	7,056	592,704
85	3,655	7,225	614,125
86	3,741	7,396	636,056
87	3,828	7,569	658,503
88	3,916	7,744	681,472
89	4,005	7,921	704,969
90	4,095	8,100	729,000
91	4,186	8,281	753,571
92	4,278	8,464	778,688
93	4,371	8,649	804,357
94	4,465	8,836	830,584
95	4,560	9,025	857,375
96	4,656	9,216	884,736

n	Triangular numbers (t_n)	Square numbers (n^2)	Cubic numbers (n^3)
97	4,753	9,409	912,673
98	4,851	9,604	941,192
99	4,950	9,801	970,299
100	5,050	10,000	1,000,000

Notice that in only twenty members of our list of one hundred is the units digit the same for each entry.

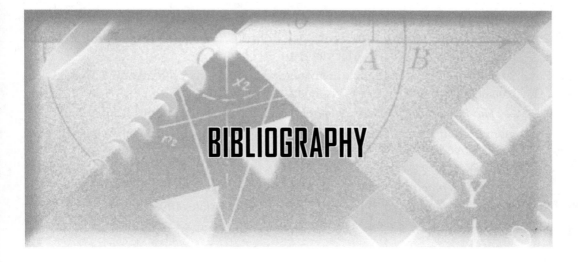

BIBLIOGRAPHY

For the reader who would like to have sources for proofs of the many fantastic relationships we encountered in these chapters, we offer the following list of books:

Alsina, Claudi, and Roger B. Nelson. *Math Made Visual: Creating Images for Understanding Mathematics*. Washington, DC: Mathematical Association of America, 2006.

Altshiller-Court, Nathan. *College Geometry. A Second Course in Plane Geometry for Colleges and Normal Schools*. New York: Barnes & Noble, 1952.

Conway, John H., and Richard K. Guy. *The Book of Numbers*. New York: Springer, 1996.

Coxeter, H. S. M. *Introduction to Geometry*. 2nd ed. New York: Wiley, 1989.

Coxeter, H. S. M., and Samuel L. Greitzer. *Geometry Revisited*. Washington, DC: Mathematical Association of America, 1967.

Dörrie, Heinrich. *100 Great Problems of Elementary Mathematics*. New York: Dover, 1993.

Gay, David. *Geometry by Discovery*. Hoboken, NJ: John Wiley & Sons, 1998.

G.-M., F. (=Frère Gabriel-Marie). *Exercices de géométrie, comprenant l'exposé des méthodes géométriques et 2000 questions résolues*. 6th ed. Paris: Editions Jacques Gabay, 1991. Reprint of the 1920 edition.

Johnson, Rober A. *Modern Geometry*. Boston: Houghton Mifflin, 1929.

Maxwell, E. A. *Geometry for Advanced Pupils*. Oxford: Clarendon, 1949.

Nelson, Roger B. *Proofs without Words*. Washington, DC: Mathematical Association of America, 1993.

——— . *Proofs without Words II*. Washington, DC: Mathematical Association of America, 2000.

Posamentier, Alfred S. *Advanced Euclidean Geometry*. Hoboken, NJ: John Wiley, 2002.

Posamentier, Alfred S., J. Houston Banks, and Robert L. Bannister. *Geometry: Its Elements and Structure*. New York: McGraw-Hill, 1977.

Posamentier, Alfred S., and Ingmar Lehmann. π: *A Biography of the World's Most Mysterious Number*. Amherst, NY: Prometheus Books, 2004.

———— . *The Fabulous Fibonacci Numbers*. Amherst, NY: Prometheus Books, 2007.

Posmentier, Alfred S., and Charles T. Salkind. *Challenging Problems in Geometry*. New York: Dover, 1988.

Pritchard, Chris, ed. *The Changing Shape of Geometry: Celebrating a Century of Geometry and Geometry Teaching*. Cambridge: Cambridge University Press, 2003.

Singh, Simon. *Fermat's Last Theorem*. London: Fourth Estate, 1998.

INDEX